United States Women in Aviation 1940–1985

Deborah G. Douglas

SMITHSONIAN INSTITUTION PRESS

Washington and London

Library of Congress Cataloging in Publication Data
Douglas, Deborah G.
United States women in aviation, 1940–1985.
Bibliography: p.
Includes index.
1. Women in aeronautics—United States. I. Title. II.
Series.
TL521.D68 1990 629.13′0082—dc20 89-600095
ISBN 0-87474-382-6

Manufactured in the United States of America
97 96 95 94 93   5 4 3 2

⊗ The paper used in this publication meets the
minimum requirements of the American National
Standard for Permanence of Paper for Printed
Library Materials Z39.48-1984

Illustrations
Front cover: Jacqueline Cochran (S.I. photo 72-6626)
Back cover: Lynn Rippelmeyer (detail of fig. 63).
Courtesy of David Hollander

D0509527

$\wp$

CONTENTS

Foreword

At the dawn of this century, when the first fragile "flying machines" were taking to the air, there was much speculation on how the phenomenon of flight might affect those who ventured aloft. Some argued that the immensity of space surrounding pilots, and the tremendous field of view that they would enjoy, would ultimately have an effect on the psyche; the mental horizon would expand with the field of vision, and the aviators would be more open-minded and receptive to what was new and different. The present work, which chronicles American women's participation in aviation with their male counterparts, offers some interesting—if challenging—commentary on the "receptivity" of the flying fraternity.

But Deborah Douglas' study offers much more: it illuminates the considerable variety of roles that American women were able to play in the development of aviation over the past half century; and the author holds these women up to the reader's view not like so many mounted specimens, but in their cultural milieu, and against the proper socio-economic backdrop. What is more, the story of women in aviation that is related in the following pages is offered as part of a larger story, one that involves us all: that of American society's struggle to adapt itself to a world ruled by technology. Then too, woven through the narrative are threads of the great social questions that have stirred our society in past decades: racial integration, the Equal Rights Amendment, the anti-war movement of the sixties, and other issues in our recent history.

Yet at bottom the story related in this book is not just about social issues, but about specific women and airplanes. The narrative is peopled with remarkable personalities such as Jacqueline Cochran and Nancy Love, whose differences of opinion had important consequences when women began to enter the military service. The reader will encounter General "Hap" Arnold, who stated plainly "women can fly as well as men"—yet the U.S. Army Air Forces, which Arnold headed, sometimes used women to demonstrate warplanes with fearsome reputations to airmen who would be flying them. The male reasoning of the 1940s could be counted on to draw the desired conclusion: if a woman would fly it, then it couldn't be all that difficult.

Then too, the story has its curious twists and turns, its sometimes strange alliances of interest. There was, for example, the unlikely convergence of anti-war feminists and anti-feminist conservatives in their mutual opposition to the training of women as military pilots. On the other hand, between male airline pilots and female flight attendants there was at first no meaningful coalition for collective bargaining purposes; it took time for their common economic interests to bridge the gender gap.

Significantly, Douglas' treatment begins on the eve of American involvement in World War II, an involvement that would prove vital for women's role in American aviation. Claudia Oakes, who traced the women's story during the preceding decade (*United States Women in Aviation, 1930-1939*), concluded that by 1939 "women were no longer oddities in any area of aviation"; she had found examples of women who were not only pilots and stewardesses, but also engineers, mechanics, and specialists in demonstrating and selling airplanes. In a sense the "heroic" age was over by 1939; the pioneers had staked out various claims to competence, but it remained for the succeeding generation to exploit those claims.

Into the early forties the movement of women into the various aspects of aviation was not very impressive, at least in the statistical sense. The Ninety-Nines, an extremely influential international association of women pilots, which figures prominently in this study, had only four hundred members in 1940; the number of stewardesses—ancestors of the modern flight attendants—was probably not much greater. But we tend to forget just what an exclusive sort of activity any kind of flying was in that period. According to the Bureau of the Census there were 130 million Americans in 1939 but only some 30,000 of them possessed a pilot's license. The nation's airfields held 13,000 aircraft, while its roads carried 30 million motor vehicles. The easily overlooked fact is that the "Golden Age" of American aviation was one in which millions were content to know the joys of flight vicariously; only a tiny fraction could fully savor the real thing.

If we leave aside military and commercial aviation, and consider flying as a diversion or a sport, then it was in Douglas' words, "a wealthy white man's sport." Like polo, it required a financial outlay that put it out of the reach of most people—and to some degree this is still true. It is not surprising, then, that the outstanding women in aviation in the 1940s and 1950s were often women of some leisure and means. Douglas demonstrates that when they took to the air they constituted something of an elite caste, stamping their avocation with their values and sometimes with their personalities. They also brought into the male-dominated world of flight a certain self-assurance, thereby winning acceptance, respect, and the acknowledgment that women pilots could be as good as the best.

At the other end of the economic scale were the stewardesses. Young and single—as was the bulk of the female labor force in the prewar era—they took jobs with the airlines for the same reason that they took sales positions in the five-and-dime stores: though not very rewarding financially, the job was a convenient way station on the road to marriage. While youth and celibacy were formal requirements, more subtle means of screening removed non-white candidates and those who were not physically attractive. An airline executive put it bluntly: his company strove to hire stewardesses who would "look good in their uniforms." Recruited and trained under such guidelines, the stewardess became, according to historian Mary Ryan, "the archetype of the sexy female in the contemporary work force."

By the middle years of World War II, Douglas finds women becoming a noticeable element on aircraft assembly lines and elsewhere within the industry, for the war had the effect of forcing many people's lives out of their habitual channels; it pulled 15 million American men from their jobs and put them into uniform, and in many cases it summoned women to take their places in the nation's shops and factories. Here Douglas has amassed statistics that are truly impressive. While in the late thirties female employees were not much more than a trace element in the plants of Boeing, Douglas, and other aircraft firms, by late 1943 women constituted 36 percent of the industry's work force, with just under a half-million women working in the industry. As Douglas points out, along with the familiar figure of Rosie the Riveter, there were women who were engineers, test pilots, and other specialists.

It has become a habit of mind among historians to regard women's expanded role in the wartime work force as an essentially temporary thing. Such studies as D'Anne Campbell's *Wives, Workers and Womanhood* have emphasized the rapidity with which American women abandoned their wartime roles to return to more traditional ones. In the aircraft industry, as in other sectors of the economy heavily committed to the war effort, the end of the fighting in 1945 saw massive reductions in the work force. Women made up a significant proportion of those released, since their sex placed them in the "last hired, first fired" category. Yet Douglas has marshalled convincing evidence showing that women remained in the industry in numbers far exceeding prewar levels. The ground that they won here during the war they were able to hold.

The same was true with the armed services. Here too during the war the iron law of necessity overrode custom and prejudice, put women in uniform, and brought them into the world of military aviation—not in token numbers but by the thousands. Women had entered the services in the previous war, but as a temporary solution to the manpower shortage; now, by 1945, they had so demonstrated their worth that they became a permanent part of the nation's military establishment. Douglas shows that acceptance did not always come easily, but it did come. All of the seemingly insuperable barriers to women flying warplanes ultimately shrank to simple problems, such as cockpit designs that ignored the physiological characteristics of women—and of small men, for that matter.

If the late 1940s and the 1950s were the time for consolidating wartime gains, then the sixties and seventies were decades in which women advanced to new milestones in the field of aviation, and Douglas meticulously points them out to her readers. Women gained admission to ROTC programs and to the service academies; they broke the no-marriage ban, which airlines had imposed on flight attendants; they followed the same training for military pilots as men did; they moved into the pilot's seat on commercial airliners.

A final chapter offers a *tour d'horizon* on the threshold of the space age, and provides a vantage point from which to look back over the changes since 1940. If the progress has been uneven—there are still very few women piloting commercial jets, for example—nevertheless, women have extended their role in the world of aviation, and in the process they have brought about some fundamental changes in attitudes. And that is as it should be—so the reader may reflect as he or she turns the final pages of this solid and satisfying book.

Lee Kennett
University of Georgia

Acknowledgments

First to be thanked are the countless women from all parts of the United States who have written letters, sent photographs, shared scrapbooks, and recorded impressions of their own experiences in aviation. At conventions and lectures, I have been besieged with offers of assistance; over dinner and hours of conversation, I have learned much, and I have come to admire the passion of these women for all things involved with flight. Truly, it has been an inspiring experience for me, and this study would not have been possible without their help.

A few individuals and organizations deserve specific mention. These include Jean Ross Howard and members of the Whirly-Girls, Inc., International Women Helicopter Pilots; Hazel Jones and members of the Ninety-Nines, International Women Pilots Association; Dorothy Niekamp, author of "Women and Flight, 1910–1978: An Annotated Bibliography," an unpublished bibliography produced by the Ninety-Nines (I was given permission to use this work extensively for my own research); Lt. Colonel Yvonne C. Pateman, USAF (Ret.); and Professor Lee Kennett of the University of Georgia.

The National Air and Space Museum (NASM) staff has been exceedingly helpful at each stage of research and writing. Claudia Oakes, Tim Wooldridge, and Von Hardesty deserve special thanks in addition to Dom Pisano, Bob van der Linden, Dorothy Cochrane, Tom Crouch, Anita Mason, Sybil Descheemaeker, Larry Wilson, Bob Drieson, Phil Edwards, Mary Pavlovich, Trish Graboske, David Romanowski, and Katie Schwartzstein. Barbara Spann of the Smithsonian Institution Press was a wonderful editor.

I would also like to acknowledge several other individuals. My friends with the Daedalus Human Powered Flight Team and Roadwork, Inc., provided much support throughout the time when I was writing and revising this monograph. The Bartletts—Mike, Tara, Jessica, Darren, and especially Carol—have always shown their enthusiasm for this project; it has been much appreciated. Likewise, my family, in particular my parents, Priscilla and Neiland Douglas, offered their love, support, and advice, for which I am eternally grateful.

Washington, D.C.
June 1988

United States Women in Aviation
1940-1985

Deborah G. Douglas

Introduction

The film flickered on a small screen, part of an exhibition on air traffic control at the Smithsonian Institution's National Air and Space Museum. Girls and boys crowded on the floor, whispering and giggling with each other. They said something about a field trip and a research question. Then the short movie began, the pencils emerged, and they started scribbling down notes on the information being presented. None seemed to notice how many women were in that film. It was hard to know which was more revolutionary: the fact that there were so many women in the film or the fact that the youngsters did not notice.

United States Women in Aviation, 1940-1985 deals with a remarkable piece of American history. It is about the aviation experiences of certain individuals who were set apart and defined as a group because of their sex. The words "women in aviation" represent an important historical construct, first articulated during the Second World War, and then actively cultivated by the aviation community until the early 1980s. This collective identification has not always existed. The first three volumes of the *Smithsonian Studies in Air and Space* series on United States Women in Aviation chronicle a very different period—an era of individual achievers. Stretching from the turn of the century to 1940, these three volumes cover the intrepid and unflappable pioneers who indicated right from the start that neither the dream of flight nor the daring required to become an active participant in the development and application of aviation was exclusively male.

1940 marks an important turning point for both women in aviation and this series. The status of women and the state of the world changed significantly with the advent of war. With these changes came a dramatic

Deborah G. Douglas, formerly Research Assistant, Department of Aeronautics, National Air and Space Museum, Smithsonian Institution, Washington, D.C. 20560.

expansion in the depth and scope of women's involvement in aviation. It is at this moment that the collective concept "women in aviation" acquired its particular distinction. An important consequence of this was that aviation came to be broadly construed as including all aspects of flight. It was acknowledged that the assumption of a pilot's primacy ignored the fact that many of the other activities (from engineering to air traffic control) could be as intellectually demanding and fulfilling as piloting.

Following the war, the position of women in aviation mirrored that of all American women. As the United States became an increasingly technological society, the means of full citizenship were necessarily linked to an individual's ability to understand and manipulate the machines, processes, and systems that shape modern existence. The study of women in aviation, in which the role of technology in history is explicit, helps reveal an important facet of the history of all American women in a constantly changing environment.

This volume is an introductory survey, highlighting the most critical events, organizations, and individuals in the war and postwar periods. It is organized chronologically and explores the experience of both civilian and military women. Their stories are told using anecdotes, statistics, and documentary evidence. There are the painful stories of discrimination based on class and race as well as gender. But there are also the triumphant tales of fabulous accomplishment. These pieces of historical data show that the history of women in aviation has not been one of straight-line progress. This melange of distinct occupations and avocations ebbs and flows with the larger currents of American history.

The activity of flight has been dynamic and dramatic, which has meant a heightened public visibility for all of its participants during the 45-year period. The women who are the subject of this book have often

1

been the focal point of vigorous public debate about critical social questions. These questions range from what the roles of men and women should be in the United States to whether or not women should serve in combat. The text attempts to identify the critical questions at important junctures of change and development.

The changes in popular attitudes, such as evidenced by the youngsters described in this Introduction's opening paragraph, constitute one of the most striking pieces of evidence that the 1980s represent a transition phase for women in aviation from one era to another. This study records the fascinating turn of events in the recent history of women in aviation which led to that moment. It also identifies the resources that will enable more detailed research to be done in specific fields of related inquiry. All of these works confirm one basic notion: the desire to fly is a universal human experience. The following chapters will help define what that has meant for American women since the beginning of World War II.

World War II

1. Students and Teachers, Clubs and Colleges: Women in Civilian Aviation Organizations

O-o-o-oh, we flew through the air with the greatest of ease,
The C-A-A's feminine C-P-Ts:
But they tell us "Men Only," they've kicked us right out.
Now it's "Pay as you pilot, pu-lease."

from *Pilette's Lament*, author unknown

In the early morning hours of the 7th of December 1941, Cornelia Fort, a young flight instructor with Andrew Flying Service, was working with a student pilot near the John Rogers Airport in Honolulu, Hawaii. Her Sunday schedule included a busy teaching program in between giving a number of aerial sightseeing tours. That first lesson of the day was typical, filled with many practice takeoffs and landings, and continuing work on the skills her student needed to qualify for his first solo flight. Suddenly, however, just before the two were about to land, Fort saw a military airplane speeding directly toward her. She grabbed the controls away from her student and managed to pull just above the oncoming airplane. Fort described what she saw through the still-shaking plastic window:

The painted red balls on the tops of the wings shone brightly in the sun. I looked again with complete and utter disbelief. Honolulu was familiar with the emblem of the Rising Sun on passenger ships but not on airplanes.
I looked quickly at Pearl Harbor and my spine tingled when I saw billowing black smoke. Still I thought hollowly it might be some kind of coincidence or maneuvers, it might be, it must be. For surely, dear God . . . .
Then I looked way up and saw the formations of silver bombers riding in. Something detached itself from an airplane and came glistening down. My eyes followed it down, down and even with knowledge pounding in my mind, my heart turned convulsively when the bomb exploded in the middle of the harbor. I knew the air was not the place for my little baby airplane and I set about landing as quickly as ever I could. A few seconds later a shadow passed over me and simultaneously bullets spattered all around me.[1]

The Japanese had attacked Pearl Harbor. The next day, 8 December, the United States declared war and formally entered the conflagration that had engulfed most of the rest of the world.

The United States had been sending huge convoys of arms and supplies to Britain and anxiously watching events in Europe and Asia, but it was not until the bombing of the American fleet at Pearl Harbor that World War II had a great impact on the daily lives of most Americans. The 1940s were a time of great change in American society. The war had the effect of unifying the American people to a great extent, bringing about a more cohesive community based on a broader and more inclusive definition of its citizenry. The necessities of war wrought greater social and economic justice but did not eliminate all of the barriers. Different groups progressed at varying rates depending on their status at the war's commencement. The mobilization of the nation's work force and resources to meet the war's intense demands was such a force for change in American attitudes that there was much public debate about the necessity of working toward a more democratic and just society.

The demands, the mobilization, and the debate obviously had implications for everyone involved in aviation. For women pilots like Cornelia Fort, the attack on Pearl Harbor put a temporary end to certain kinds of flying activity; it also meant the creation of a whole range of other opportunities. Women's flying organizations, like the Women Flyers of America and the Ninety-Nines, altered their agendas to provide wartime service. Others, such as the disaster relief group, Relief Wings, Inc., originated as a direct response to the war.

The government, at both federal and state levels, expanded the number of aviation programs open to women. Women were part of the Civilian Pilot Training Program (CPTP) almost from its inception, and later many participated in the Civil Air Patrol (CAP).

3

Through the Civil Aeronautics Administration (CAA), women became Link trainer instructors,[2] parachute riggers, and air traffic controllers, professions that demanded special aviation knowledge.

The demarcation lines between the civilian and military worlds were obviously much less distinct during World War II than during peacetime. Anyone with the ability to pilot an airplane had a valuable skill, but the avenues for wartime service were less open to women than to men. Barred from combat flying, women took advantage of many aviation occupations that were newly created (or newly open to women) but which did not involve any actual flying. This resulted in a more complete integration of women into all aspects of aviation, due, not to planned development, but to the greater abundance of nonflying jobs and the shortage of qualified people to fill them.

At the time the United States became involved in World War II, the Ninety-Nines, International Women Pilots Association, founded in 1929, was one of the aviation organizations working hard, both to create new flying possibilities and to remove the restrictions imposed on women flyers. It had become a strong network by 1940, with more than 400 women pilots, whose purpose was to provide "good fellowship, jobs, and a central office and files on women in aviation."[3] In general these women tended to come from affluent or middle-class families and were well educated and white. Despite their position of economic and social privilege in American society, the leadership and inspiration these women provided for hundreds of other women in the field (and thousands who were spectators) was very important. They did not accept existing social norms concerning women in aviation; they flew in spite of them, abandoning prescriptions of proper feminine behavior without being overwhelmed by the combined disapproval of family and society for "inappropriate activity." These young women were described sympathetically by Julietta K. Arthur:

What chance for a career is offered them in that most glamourous of pursuits, the business of flying? Everybody knows that the late Amelia Earhart swiftly followed on the heels of Lindbergh across the Atlantic; that Ruth Nichols took the altitude record from Colonel Chamberlain in the same ship that he had flown; that Phoebe Omelie bettered the record of men she competed with in cross-country and closed-course racing; and that Jacqueline Cochran, now America's foremost aviatrix, is winner of more awards than she has wall-space on which to hang them. But these are the stars, the top notchers, whose dazzling careers in relation to everyday jobs in aviation are on par with Metropolitan Opera singers, to whom the hopeful aspirants of the chorus look for inspiration .... A large number of women learn to fly for the same reason they enjoy driving a swift automobile, but after they have once learned, a number have found a hobby turning into a paying pursuit.[4]

In 1939, Betty Huyler Gillies, one of the charter members, was elected president of the Ninety-Nines. Like her predecessors, Gillies developed a strong program for the organization. She worked with Alma Harwood, who was a fellow member and the prime instigator for the creation of the Ninety-Nines' Amelia Earhart Memorial Scholarship Fund. Gillies led a battle against the Civil Aeronautics Authority (the CAA would shortly become the Civil Aeronautics Administration) over regulations concerning pregnancy. In March of 1940, it was reported that Gillies, as president of the Ninety-Nines, had submitted a petition challenging the CAA's right to ban pregnant women from flying. This immediately provoked a response from CAA officials saying that while CAA medical experts disliked the practice, there was no official ban on pregnant women pilots, but there was a restriction for a short period during a woman's "recovery." If a woman's license expired during this time, then she was required both to rewrite and re-fly the CAA examinations to win back her pilot's license. The Ninety-Nines under Gillies leadership were successful in having these regulations modified as well as in exposing the exaggerated view of the dangers to a pregnant woman in flying.[5]

In 1941 the famous pilot Jacqueline Cochran became president of the Ninety-Nines. Under Cochran's leadership the organization began to respond to the new climate for aeronautical activities. The United States' entry into the war curtailed civilian flying dramatically. In many coastal regions pleasure flying was prohibited. Despite the limitations, opportunities for women actually increased. Affluent women pilots with extensive flying experience, who once related chiefly to exclusive aviation country clubs, found ways to continue to fly that included service to their country. For the first time, learning to fly became an affordable possibility for women of a lower income level because of programs sponsored by the Women Flyers of America and the CAA.

The Women Flyers of America, Inc., (WFA) was essentially a national flying club, open to any girl or woman involved in, or just interested in, aviation. The organization first invited members who wanted to learn to fly to take ground school classes (all courses offered were CAA approved). Upon completion of the course, a representative of the WFA would negotiate fees on behalf of all the women who wanted to continue with actual flying lessons. The large number of women involved meant that the representative had considerable bargaining power. The net result was to make flight lessons available at low rates. On average, WFA members paid $275 for ground instruction, dual and solo training (from CAA licensed instructors), earning a private pilot's license in 35 to 50 hours. This was about 20 percent less than comparable fees (approximately $345) for a similar program taken by an individual. It was still a "pay-as-you-go" system, but the WFA acted as the guarantor for the airfields and their instructors.[6]

The WFA's program was a remarkably popular concept. The organization, founded in April 1940 in New York, had a rocky start because the response to its modest announcement was so overwhelming. Nearly 1,000 women wrote back, two-thirds of whom enclosed the $5 membership fee. The founders, Opal Kunz[7] and Chelle Janis, had initially intended to run a trial program of "installment plan" flying in the New York City area. But almost instantly it became a national organization and the resultant mismanagement almost killed the fledgling group. Fortunately, in August 1940, one well-known WFA member, Vita Roth, a 1920s women's parachute-jumping record holder, decided to assume the responsibility for organizing the group.

Roth found that most WFA members were young (averaging 20 to 35 years old), working (earning between $18 and $75 per week), adventurous, and concerned about national defense. Indeed, many WFA members voiced the wish to be available if America had an urgent need for ambulance and liaison flyers. One representative WFA class included individuals like Eleanor Scully, the fashion editor of *Vogue* magazine, and debutante Mary Steele, as well as several secretaries, file clerks, and nurses.[8]

WFA chapters were founded in Washington, D.C., Hartford, and Philadelphia. By the time of the attack on Pearl Harbor there were 10 chapters in 10 major cities. Pearl Harbor and the declaration of war created a surge of enthusiasm for flying, and many WFA trainees were particularly interested in having the WFA set up a joint application program with the Civil Air Patrol. Programs for teaching women to fly received positive reactions from another quarter: instructors of WFA students found them easier to teach than men because they were much more inclined to listen.[9]

Ironically, director Roth was, at first, publicly opposed to women's involvement in the war, but she continued to lobby hard in Washington with defense agencies to provide opportunities for her fellow members.[10] At first she proclaimed: "We don't ever expect to be in combat: that's a job for men. It's all 'bunk' to say women are better pilots than men. We aren't. Most women don't have the mechanical talent or physical stamina of average men pilots; but we are qualified for secondary service behind the lines."[11] Immediately following the creation of the Women's Airforce Service Pilots (WASP) organization, however, Roth issued a more flattering statement, this time focusing on the abilities of leaders Nancy Love and Jacqueline Cochran: "We believe the choice of these two women is a particularly happy one and I speak for all members of Women Flyers of America when I say we stand behind them 100 percent. Mrs. Love is not only one of our best pilots in the air, she is an efficient and level-headed woman on the ground. Miss Cochran's qualifications need no elaboration."[12]

The WFA was not the only means for a woman to learn how to fly, but it was popular because of its lack of membership restrictions. Anyone could belong, whereas the Ninety-Nines was restricted to women pilots, and the Civilian Pilot Training Program took only college-age students, allotting only three percent of the slots to women. Furthermore, the WFA organization adapted well to the various demands and expectations of war. The WFA saw itself as the facilitator for women who wanted to get into the many facets of aviation, so it adopted training programs that would qualify graduates for Link trainer schooling, air traffic control school, meteorology training, and parachute rigging. It also sponsored flying lessons and flight time for those who wanted to become ferry pilots or work with the Civil Air Patrol.[13]

Another significant nonfederal program was developed in 1941 by Stephens College in Columbia, Missouri.[14] "Air-mindedness" was the theme of the comprehensive curriculum, the first such program ever offered to women, according to the president, Dr. James M. Wood. It was a cooperative venture between this women's college, the Army Air Forces, the CAA, and 12 leading airlines including Transcontinental and Western Air, Chicago and Southern, Mid-Continent, and Braniff. Stephens offered specific vocational training in many areas of aviation, always emphasizing safety and utility. There were courses for ticket clerks, reservations clerks, flight attendants, and other passenger service personnel. Stephens prepared drafting and blueprint readers and engineering draftsmen for the aircraft industry. Further, there were courses to prepare meteorologists and junior weather observers, Link trainer operators, control tower operators, flight instructors, and mechanics. The goal of the college was to make it possible for each student to at least make a flight as a passenger whether or not she wanted to learn to fly. During the first years of the program in World War II, however, this "flight experience" was a requirement only in the elementary aeronautics courses and the airline job-preparation classes.

The program was experimental but very successful. Throughout the course of the war hundreds of Stephens graduates moved into a wide variety of aviation occupations. These women were the tangible results of the vision of the college officers who believed in a future

6

aerial age accessible to all Americans.

Most college-age women (between the ages of 18 and 22) who learned to fly during World War II learned under the CAA's Civilian Pilot Training Program (CPTP). The program was conceived by CAA administrator Robert H. Hinckley in 1938 as a way of introducing American youth to the air age. For the purposes of persuading Congress, which harbored strong isolationist sentiment, the CPTP was billed as a civilian program open to young men and women as a way of stimulating civilian aircraft sales in the faltering American aviation industry. Hinckley, however, viewed the program as an important way to increase the pool of qualified pilots available for national defense. This second reason was played down and, in December 1938, President Roosevelt announced the establishment of an experimental CPTP. One year later, the program had trained 9,350 men and women at 435 colleges. Arrangements were also made to open the program to nonenrolled college-age youth through a competitive examination program.[15]

Women were generally admitted into the CPTP at the ratio of one woman to ten men. However, several women's colleges did take part in the CPTP, thereby increasing the total number of female participants. These schools included Lake Erie College, Adelphi College, Mills College, and Florida State College for Women. The CAA considered the inclusion of women to be experimental, a means for the agency to gain experience in encouraging women to become involved with general aviation. Its attitude was based on the premise that women made most of the important financial decisions in a household, and thus could have a substantial impact on the aviation sales market.

FIGURE 1.—Delphine Bohn in the cabin of a Piper Cub. She was a flight instructor for the Civilian Pilot Training Program (CPTP) in Amarillo, Texas. Bohn left CPTP to become one of the "Original 27" members of the Women's Auxiliary Ferry Squadron (WAFS). (Courtesy of Delphine Bohn, S.I. photo 86-5614)

For example, a woman who knew about aviation would be supportive of a pilot husband who wanted to buy an airplane.[16]

The CPTP was designed to be a cooperative program run by colleges in conjunction with local flying-field operators. The colleges would provide ground-school instruction (with CAA-approved teachers) in aerodynamics, meteorology, navigation, parachutes, air regulations, basic mechanics, and air history; the flight instructors would provide the elementary flying course, which included 35 to 50 hours of dual instruction and controlled solo flying. The federal government would set and maintain the standards of instruction, certifying both ground and flight personnel, and paying the expenses.[17] By the end of the program's first year there were an estimated 980 new pilots.

As war became imminent, the CPTP acquired a more military character. Early in 1941, trainees signed nonbinding agreements to serve in the military in case of war or national emergency. Even women had to fill out cards specifying the service they preferred.[18] The pledge soon became a legal obligation for enlistment in the armed forces, and thus, in June 1941, women were automatically excluded from the CPTP. This action was protested by many women, each of whom received a letter signed by Robert Hinkley, which stated:

If, or when, the time comes when trained girls are needed in non-combat work to release men for active duty, that will be a different situation. I believe that we may also assume that upon the cessation of the emergency, we would resume our former policy toward women.[19]

Eleanor Roosevelt, a strong supporter of women in aviation, publicly demanded an explanation. Though she received a personal letter from CAA officials, the essential message was the same:

It is generally recognized that male pilots have a wider and more varied potential usefulness to the armed forces than female pilots. We have had to make changes with this in mind, and during the past year our program has of necessity been closely integrated, with the needs of the Army and Navy aviation constantly in the foreground. To limit our training to young men is only another in a series of steps based on this philosophy.[20]

The CPTP officials did not want to abandon women altogether. The draft did not discriminate or differentiate in the slightest between the program's male flight instructors and their students. Both were equally likely to be drafted. This fact, combined with the rapid expansion of the CPTP, which trained 10,281 students in 1940 and 215,676 in 1944, meant there was a desperate need for flight instructors. The CPTP, and later the War Training Service (WTS), did use some women instructors, although records indicate that no more than 50 were ever employed and that their widespread use in this capacity was never seriously considered.[21]

Gertrude Meserve (now Tubbs),[22] for example, taught nine classes in Boston and Norwood, Massachusetts, as part of the CPTP 1939 test program. Later, when operations were moved inland to Orange, Massachusetts (flying was prohibited in coastal regions during the war), she taught students from the Massachusetts Institute of Technology, Harvard, Northeastern, and Tufts universities. Meserve was a flying enthusiast. Her first airplane ride was in 1935 and her first lesson was at Logan Field (now Logan International Airport) in Boston in April 1938 just prior to her high school graduation. One year later she received her commercial and instructor's ratings, and in 1941 she was appointed Flight Inspector for Private Licenses by the CAA.

Ironically, Meserve, immune from the draft by virtue of her gender, was nevertheless called up for the war effort. In September 1942 she received a telegram from Nancy Love asking her to join the Women's Auxiliary Ferry Squadron (WAFS). Meserve decided to become a member of the WAFS because the job sounded more exciting and more directly connected to the war effort than any other available opportunity to use her expertise.

The CAA made its first proposal at about this time for women flight instructors to train Army and Navy pilots. The CAA proposal was for a two-part program. They wanted to hire instructors such as Meserve immediately, and they wanted to start a training program specifically to prepare women to be flight instructors. This training program was directly inspired by the outstanding success of a state-run program in Tennessee developed by Phoebe Omelie.

Phoebe Omelie, one of America's top women pilots in the 1920s and 1930s, developed a program for training women flight instructors. In the summer of 1941, the Tennessee Bureau of Aeronautics opened a special school under Omelie's supervision funded by revenues from the state's aviation gasoline tax. The original idea had been proposed by W. Percy McDonald, chairman of the Tennessee Bureau, but Omelie deserves full credit for its successful implementation.[23] She had been in Washington during the 1930s working with what was then the National Air Marking program[24] of the Bureau of Air Commerce, but she decided to respond to McDonald's call to return to Tennessee and head up this new venture. One of the motivating factors was the strong sentiment at the CAA that flight instructing was the best way for women to be involved in wartime aviation.[25] Omelie addressed that sentiment when she explained why it was that women in other positions such as pilots or mechanics got greater news attention:

FIGURE 2.—Phoebe Omelie was an active pilot in the 1920s and 1930s, when she began working with Blanche Noyes in the National Air Marking Program. In 1941 she developed, for the state of Tennessee, a wartime program to train women to be flight instructors. (Photo from Time-Life Collection, National Air and Space Museum)

the novelty of women in those positions made good news copy in the minds of many men editors. The lack of publicity for women flight instructors did not, however, diminish the importance of their work; in fact, it indicated a higher degree of social acceptance.

Naturally, the dashing uniformed woman pilot of a roaring warplane is going to get more attention (and photography) than her flight instructor sister, just as the uniformed volunteer gets more publicity than her coveralled sister at the factory workbench. Just as naturally, however, women are finding their proper places as instructors, the position women have filled better than men for generations.[26]

The program created by Omelie was designed to start out slowly. The first session involved training only 10 women. If successful, the course would then expand as much as funding and interest allowed. An overwhelming 1,000 applications arrived, and Omelie set up a special selections committee to choose the first 10 candidates.

All of the applicants were required to have a private pilot's license with 120 or more hours of flying time.

The women had to be single or, if married, could not have dependents and their husbands had to be in active military service. Finally, they had to agree to instruct wherever and whenever the bureau decided to place them. The 12-week program was a rigorous eight-hours-per-day, six-days-per-week course that provided uniforms, food, and housing. It offered considerable benefits to participants. With more than 60 hours of flying, 200 hours of ground school, and 160 hours of flight instructor's ground school, each woman was qualified for a CAA ground instructor's rating in meteorology, aircraft structure, aircraft engines, aerial navigation, and civil air regulations in addition to becoming an "airborne" flight instructor.[27]

Graduates of the program were in very high demand. By August 1943, four months after the first class graduated, CAA officials were beseeching the Senate Appropriations Committee for $2.5 million to train 500 more women in similar programs around the country. In addition to the 500 prospective graduates, 400 more women would have at least started training during that same six-month period.[28] The request was turned down, yet the CAA continued to support other efforts on behalf of women flight instructors.

One such effort was to involve as many women flight instructors as possible in the re-rating process. Re-rating was a program instituted by the CAA just prior to World War II that required all instructors, regardless of experience, to turn in their old license and take an examination to requalify for a new instructor's rating. The purpose was to establish uniform standards for CAA flight instructors. Because there was no discrimination between men and women in regard to the examination, re-rating, by providing a tangible index of ability, helped raise the status of women within the flight instructor profession. In other words, now that there was a set of standardized test scores available for open comparison, it was quite difficult for an operator of a flying school to mask sex discrimination by claiming a woman who had participated in re-rating was not as qualified or well trained as male applicants.[29]

In the fall of 1941, 43 women were listed as instructors in CAA-qualified flight-training schools around the United States. Eventually, many more women joined their ranks. Of the original 43, most exuded great personal confidence in their abilities. Yet, they very often used the traditional ploy of an earlier generation of women pilots—"If I can fly, certainly you, a man, ought to be able to"—in order to win over potential but skeptical male students.

At the same time, these women also tended to be a little more critical of women students, whom they viewed as more hesitant. Evelyn Kilgore, a flight instructor at Tri-City Airport in southern California, explained this attitude:

Girls make good flyers. They learn slower because they don't understand the mechanical end of flying the way men do. But they are smoother and more careful. Men like to "kick a ship around." Sometimes they get a little foolish. Women respect a plane more, feel their way into the business of flying better.[30]

Interestingly enough, Kilgore, who was an outstanding pilot, contradicted her own stereotypes. She soloed after only eight hours of instruction and had a great personal commitment to aviation, which contradicted her comment that

women don't get as far as men do because they fritter around. They have to spend money and time on clothes and cosmetics and things like that. They just about get started and they fall in love, too .... Men are different. When they start flying they stick to it. If they have a girl, they bring the girl to the field. Pretty soon she's flying, too—sometimes.

Most women instructors only rarely made generalizations about gender differences in learning to fly. Betty Martin, a CPTP graduate who got her instructor's rating and then taught in Texas, was quick to identify the real challenge of flight instruction, which was (and is) to find the right approach for each student. Many of the people involved with aviation during World War II were convinced that women made better instructors because they believed women were natural teachers with an "inherent ability to put themselves in the student's place."[31]

Flight instruction was not the only means of civilian participation in the war effort, but it was one of the few paying options for women. A small number of women operated aircraft service centers at small local airports. Called fixed-base operators, the women and men who owned these businesses provided such services as maintenance, fuel, and preflight preparation. Many also provided flight instruction or at least served as the contact point for students looking for teachers.

For women who did not require remuneration, there were several volunteer aviation organizations in which they could be involved; for example, a woman might serve as an Air Raid Warden or Aircraft Spotter for the Office of Civilian Defense. One such volunteer group, founded just prior to the American declaration of war, was Relief Wings.[32] Conceived by Ruth Nichols, an outstanding pilot of the interwar years, Relief Wings was an air-rescue service comprised of private airplanes, volunteer medical professionals, and a network of medical facilities. Through careful organization of pilots, aircraft, doctors, nurses, airports, and hospitals, the organization was an additional component in a complex program of civilian defense, able to provide high quality air-ambulance service. Thus its slogan was "Humanitarian Service by Air," and it received the active support of many airline executives.[33] Organized and run by women, all of its positions were also fully open to female participation. The creation of a well-organized group with a firm financial base took much of 1941, but when the Japanese attack shook the nation, Relief Wings was ready and able to function across the United States. In order to solve the problems of restricted private flying, Nichols went to Washington to confer with the appropriate military and aviation officials. The result of these meetings was to make Relief Wings an adjunct service of the Civil Air Patrol.

Organized by the Office of Civilian Defense just before the Pearl Harbor attack, the Civil Air Patrol (CAP) was another group that provided opportunities for women in aviation during the war. In the first months of 1942, the details of how the CAP would be structured were released. The stated purpose was to "weld civil airmen—and women—into a force for national defense by increasing knowledge and skill in every type of aviation activity."[34] The CAP would perform important courier services, coast and forest patrol, and ferrying operations within the United States, in order to release aviation personnel of the Army and Navy for active war service.[35]

The CAP was particularly receptive to teenagers and women. Women, however, were not permitted to fly coastal patrols, even though they often instructed the men who would. For example, Dorothy Heberding, a CAP flying instructor, taught antisubmarine patrol duty at a Florida base. She was a master sergeant, yet she herself was not permitted to join the aerial lookout for German submarines.[36] In an interview with *Flying* magazine,[37] Heberding commented that her pet annoyance was "the reluctance of people generally to accept a woman whether as a pilot or a preflight inspector." She frequently would hear, "Hhh! A fee-male pilot to check us out! Nothing doing."

Because the CAP desperately needed pilots, however, gender was frequently overlooked. Women and girls who wanted to get involved felt a certain pressure to do well. Georgette Chapelle put it like this when writing about the CAP:

We women are trained to companionship, not generalship; to discussion, not discipline; to compromise rather than the "greatest good for the greatest number" policy by which a semi-military organization works.

Put it this way: Women are accepted in the patrols as equals with men—in flying, in leadership, in the suppression of temperament. It's up to you to prove that this is the basis on which you want to remain, for every girl who accepts discipline cheerfully makes it easier for the next ten girls to follow her pioneering.[38]

Chapelle's statement underlines the lack of restrictions applied to participation in the Civil Air Patrol. Yet, given the volunteer status of the task coupled with the relatively large time demand, it is also obvious that participants needed to be free from such obligations as full-time childcare. In addition to being open to women, the CAP did not discriminate on the basis of race. The well-known black aviator of the 1930s, Willa Brown, was active in the CAP. As a lieutenant and adjutant, she was the first black woman to be an officer in the organization. Based in her native Chicago, Brown taught aviation courses in the high schools and organized a CAP squadron. Her impact on aviation in World War II, however, went far beyond her CAP activities. She was the coordinator of war-training service for the CAA, and more importantly, was the director of the Coffey School of Aeronautics, the school selected by the Army and the CAA to "conduct the experiments" that resulted in the admission of blacks into the Army Air Forces. Later, Coffey became a feeder school for the Army Air Forces' program for black aviators at Tuskegee Institute.[39]

Willa Brown's work as a teacher of aviation classes is a good example of the type of nonflying activities that were part of the Civil Air Patrol. Other tasks included maintenance, radio operations, weather forecasting, chauffeuring, and first aid work.[40]

The U.S. Weather Bureau offered graduate scholarships in meteorology to both men and women who were college graduates (with a year of calculus and a year of physics) and holders of private pilot's licenses.[41] Training involved an eight-month course that resulted in the professional rating of junior meteorologist, earning about $2000, plus overtime, per year. Very few women became meteorologists (by 1948 only eight women were professional meteorologists), but one important woman in this group was Lois Coots Tonkin, a graduate of Marietta College in Ohio, who learned to fly with that college's CPTP program. Inspired by her instructor, Lenore Harper McElroy, Tonkin advanced in the CPTP until she became a ground school instructor. A college physics professor heard of the CAA meteorology program and called Tonkin to see if she was interested. As a result, she became the only woman in a class of 200 at New York University and eventually the first woman metorologist at the Weather Bureau in Washington. Later Tonkin worked in Denver and then Detroit. In each place she encountered an initial resistance toward women, only to see that prejudice melt away as the men adapted to her presence.[42]

Women also worked for the Civil Aeronautics Administration. Like the Civilian Pilot Training Program, the CAA also experienced rapid depletion of its personnel ranks precisely when its need was increasing. One of its solutions was to develop intensive training courses for which the CAA encouraged inexperienced

workers to apply. Another program was launched specifically to encourage women to apply for positions with the CAA. To enter at the more advanced level, one qualification that could be substituted for technical experience was a pilot's license. This allowed some women to begin their career with greater administrative responsibility and higher pay than was typical for women in the CAA at the time.

Most women interested in aviation, however, were encouraged by the CAA to start as aircraft communicators. These persons were responsible for taking and reporting weather observations and operating and maintaining radio telegraph, radio telephone, and teletype equipment. Applicants had to be between 17 and 40 years old and able to type 40 words per minute to qualify for a trainee position. Once the orientation and training program was successfully completed, the trainees would become junior aircraft communicators. It was noted by a CAA personnel officer, Edward J. Gardner, that in 1942, 75 percent of the trainees were women.[43]

The CAA also trained supervisors in fields of aviation-related instruction that were controlled by the federal government. Maintenance supervisors oversaw the process of training and certifying aircraft and aircraft engine mechanics. Likewise, CAA ground school supervisors and flight supervisors were responsible for instruction, training, and certification in their respective fields.

In a different section of the CAA was the Air Marking Division under the leadership of Blanche Noyes. Air marking involved making signs that would be clearly visible to pilots and would indicate location and compass orientation. Started with a few women in 1935 under the Bureau of Air Commerce, it had been the first United States government program conceived, planned, and directed entirely by women. Owing to the war and the fear that air marking would aid enemy aircraft, the division reversed its task and began a program to remove all of the signs.[44] At the war's end, it would revert back to its original function.

The job for which the CAA was best known was air traffic control. This was still a relatively new and developing field. There were a multitude of openings because the CAA had recently taken over control of virtually all airport towers (many had previously been under local or state jurisdiction).[45]

To staff the towers the CAA developed an intensive training program, which was taught at seven centers located around the country. According to CAA estimates, about one third of the controllers were women. It was specifically noted that their salaries were identical to those of men in equivalent posts.[46] Together, these men and women helped create the air traffic control system that is essentially the one employed today. It also represents one of the most immediate and visible links between the federal government and the air transportation industry.

The jobs within the CAA had identical counterparts in the military, and because of the war, the distinction between military and civilian responsibilities and occupations was blurred. Organizations such as the CAP took on a military air while others, such as the Women's Air Force Service Pilots, though technically having civilian status, truly became in thought, word, and deed part of the military. Other aspects of aviation, such as air transportation and industry, had intimate ties to the military and the government and yet maintained a thoroughly civilian orientation. The next chapter explores these important segments of the aviation community, broadening the role of women in aviation from primarily avocational and unremunerated service roles to vocational (and therefore compensated) positions.

Many of the individual women and groups discussed in this chapter could never have been selected for an elite military program; they were disqualified because of marriage or children. Others did not choose to seek paid employment. The organizations such as the Civil Air Patrol and the Ninety-Nines, the college programs, and the Civilian Pilot Training Program provided these women with their only opportunities to begin or continue their involvement with aviation during the wartime years. Their contribution is an integral part of the larger history of women in aviation.

2. Coffee, Grease, Blueprints, and Rivets: Women at Work in the Aviation Industry

We are beginning to find that there is no work dependent upon skill and dexterity that women cannot be taught to do very well.

Fairchild Aircraft Company Official

The women who worked in the control towers or in other capacities for the Civil Aeronautics Administration remind us that women have been employed in a variety of positions in the air transportation business for a long time. From the earliest days of aviation, women had been involved with selling flight. During World War II the professional establishment of the flight attendants led airlines to experiment with women in other occupations, such as maintenance.

The demands of the war led to a huge expansion in the aircraft industry and enormously enlarged the opportunities in it for female employment. From engineer to riveter, women in the industry assumed new roles and consequently, like most women war workers, acquired a new view of their abilities and of their place in American society.

Prior to the war, the most prominent women in commercial aviation were those working with the airlines as flight attendants, ticket agents, reservations clerks, and other service personnel. Although two airlines continued to use male stewards (Eastern and Pan American), the draft promptly insured women's virtual domination of the positions. Women did not have final authority over the processes of selection, training, and evaluation, although some flight attendants eventually became "chief hostesses," a position in which they exercised some managerial responsibility.

The ultimate airline position for women, the one which before the war personified women in aviation for most people, was the flight attendant, or stewardess (or hostess as she was then called). In 1940 the standard requirements for each applicant for a hostess position were that she had to be between 21 and 26 years of age; between five feet, two inches, and five feet, six inches tall; between 100 and 125 pounds (weight had to be proportionate to height); a graduate registered nurse; a United States citizen; single or widowed; personable; intelligent; and attractive. These qualifications are apparent in Braniff Airlines' advertisement for their hostess positions:

Applicant must be of perfect physical condition, well proportioned, no disfigurations. Unquestionable family background. Irreproachable character. Poise, self-confidence, tact, diplomacy. Pleasant personality. Ability to deal with people.[1]

What these requirements mask is the discrimination against black and other minority women, which mirrored the character of the American aviation community in general. The airlines had a very specific concept to sell to the public, and neither the airlines nor the American public confronted the social dictates that limited the applicant pool to white middle-class women.

Airline managers believed that most of their patrons would be white middle-class men and that it would be "difficult for potential travelers [especially men] to admit fears of flying when young women routinely took to the air as part of an inflight crew."[2] The restricted nature of the pool of hostesses was not questioned or even noticed by most passengers or by the airline corporate executives. As Alice H. Cook puts it in her introduction to Georgia Panter Nielson's book on the history of the Association of Flight Attendants union:

From the beginning the flight attendants were treated like daughters of Victorian, middle-class families, girls who needed protection during the few months they would work. It was expected, of course, that they would not question the decisions of their employers or of the captains in the cockpits, and that they would just as unquestioningly stop work after they married.[3]

Marriage was a dilemma for the airline managements. It was to their advantage to maintain a high turnover rate, because companies did not have to have elaborate wage increases, establish pension plans, or maintain expensive benefit programs. In fact, the base wage (approximately $125 per month) for attendants remained almost constant from the time they were first hired in 1930 through the end of World War II.[4] Further, the administrators liked to think of their "girls" as a highly desirable lot. For example, one Transcontinental and Western Air (later Trans World Airlines) instructor reportedly advised his class: "If you have not found a man to keep you by the time you're 28, TWA won't want you either!"[5]

FIGURE 3.—Flight attendants from Challenger Airlines (a Frontier antecedent) are shown posing in full uniform for a public relations photograph. Until World War II all attendants had to be registered nurses. (Courtesy of the Association of Flight Attendants, S.I. photo 86-11820)

The catch in this situation was that the airlines had made a financial investment in training the women for the job, an expensive proposition, when they might never work for the company. Hence the rule evolved that an attendant had to work at least a year before leaving, unless her future husband (or anyone else for that matter) was willing to "buy her freedom."[6]

World War II changed the flight attendant profession. By 1940 the Douglas DC-3, the most widely used civilian aircraft, and, indeed the whole concept of air travel it helped establish, had become fixtures in the mind of the air-conscious public. Carrying 21 passengers, the DC-3 expanded the airline passenger population and made commercial airlines highly profitable. The new generation of air passengers had high expectations of in-flight service, including meals as well as a multitude of other amenities. The DC-3, with its extended range and larger passenger load, represented a substantive increase in the workload for the hostess.[7] The airlines could not offer these more sophisticated services if, as had largely been the case until the DC-3, they depended exclusively on the copilot to interact with the passengers. This led to hiring substantially more women as hostesses.

In the 1930s, to justify the resulting extra weight and reduction in revenue passenger space occasioned by the addition of hostesses, the early promoters of this female flight-crew position had decided that the women should be registered nurses. They believed the nurse/hostess would provide psychological reassurance through the image of safety and caution. The nursing requirement was dropped in the early 1940s as military and civilian hospitals began to experience a desperate shortage of nurses.

Although this change in qualifications opened the door for more applicants, it must be admitted that the nursing requirement had lent a degree of prestige and status to the occupation. A trained nurse was a professional, and work as a stewardess represented an unusual adventure and a challenge. It might lead to marriage, but if not, a woman's nursing training would enable her to provide for herself. Ellen Church, the first woman hired as a flight attendant, is an excellent example. Church worked for United Airlines for a year and a half until she injured her foot in a car accident (her tenure was a typical period of service for attendants). She then returned to nursing and in 1942 entered the Air Evacuation Service of the Army Nurse Corps. In September 1944 she became the first flying nurse to be awarded the Air Medal.[8]

Ellen Church, and many women like her, had a strong sense of self-esteem. She was intelligent, well organized, and highly motivated; she was also extraordinarily well trained for the nature of the job that the airlines had in mind. Hostesses were supposed to be "elite temporary workers [in a] . . . short-term position for women to enjoy before being married to raise a family."[9]

The new women hired from 1940 to 1945 recognized the disparity between the increased workload and the lack of a subsequent increase in pay. For example, at United Airlines the base pay in 1942 was $125 per month. If a woman continued with United, after six months her pay would be increased to $140 per month. During a second year of employment her pay would be $150 per month. Subsequently her wage would be increased annually by $5 per month until a maximum of $170 was reached. Despite the white collar status of the flight attendant, the base salary was comparable to the wage of airframe production-line workers (who earned about $.96 per hour on average). However, the stewardesses usually worked only 100 to 115 hours per month compared with the minimum of 160 hours for production line workers.[10]

During the war some stewardesses were required to work as many as 160 hours per month. This is not surprising as the number of revenue passenger miles on United States domestic airlines increased approximately 220 percent from 1,052 million in 1940 to 3,362 million in 1945.[11] It was a difficult, if not impossible, time for the women to raise the issue of maximum hours and minimum wages with the airlines. Only 12 out of every 2,000 applicants were selected for training and employment. The intense competition for positions meant that the airlines always had an adequate labor supply, despite the war. This, combined with the no-marriage rule,[12] meant that the flight attendants did not initially develop any professional associations or unions, in contrast to the successful Airline Pilots Association.

By 1944 the attendant's job had assumed a prestigious air. A United Airlines lawyer of that period later recalled: "The stewardess was an important part of our culture. This was not a working woman; this was a glamour job."[13] That attitude was reinforced by the media. It was also true that when compared with an assembly line position or secretarial work, the attendant's assignments were more diverse. However, as a result of their wartime experience, the women began to reexamine their occupation. They recognized that they were making a contribution to the war effort, and most stewardesses, unlike corporate airline officials, definitely perceived themselves as working women. It was in the mid-1940s that highly motivated individuals such as Ada Brown began to emerge from the ranks of the profession. She had become a stewardess with United in 1940, and in 1944 she began the movement that led to the unionization of the flight attendants.

FIGURE 4.—Ada Brown worked as a flight attendant during the early 1940s. She became the chief stewardess for United Airlines in 1943, but returned to her former position in 1944 in order to organize the first flight attendants union, the Air Line Stewardess Association (ALSA). (Courtesy of the Association of Flight Attendants, S.I. photo 86-11861)

Before boarding the airplane the passenger's first contact was with the "voice with a smile," or the "hello girl," as reservations clerks were commonly called. The job involved booking reservations and confirming tickets, a task made infinitely more complex in those days by the lack of computerization and by the particular demands of wartime. The war forged a special relationship between the federal government and the airlines. Seats were automatically given to the military and individuals traveling on war-related business. Further, the airlines assisted the military by providing air transportation around the globe.

The reservations clerk, as was typical of many other airline positions, was required to have qualifications at a level that far exceeded the actual demands of the job. A college graduate with experience in sales or secretarial work was preferred. In addition, it was stated by the airlines that women should be "unafraid of figures, for a 'consolidated timetable,' showing international airline, bus and train schedules, is the bible of their work. They [should be] alert and steady-nerved, accustomed to handling a multitude of details under pressure."[14]

Both the "hello girls" and the women hired to work as ticket agents at public counters attended a six-week initial training course. The graduates often called this experience "personality training," but in addition to learning to alter their speech habits and understand customer psychology, they also studied timetables, airline routes, and common law as it applied to commercial transportation. Although these technical subjects represented the intellectual substance of the position, it was the cultivation and expression of feminine "charm and grace" that mattered to the companies and the clientele, as the "hello girl" image indicates.[15]

Not all women airline workers at this time were flight attendants or ticket handlers. Many became involved in the process of maintaining the airlines' fleets. Servicing the huge number of aircraft required the same degree of job specialization for mechanics as did the production line. New wartime mechanics were assigned to specific tasks rather than assuming responsibility for the entire aircraft. A small crew of specialists could service many more airplanes than the same number of individuals working alone. The other advantage of specialization was that it simplified the training process for new workers. Women who generally had had little or no experience in the field could be rapidly integrated into the hanger crews if they had to master only one facet of the occupation. Thus, the war produced teams of technicians, each masters of a small field, replacing the more traditional image of a multi-skilled—and masculine—mechanic.

Transcontinental and Western Air started a program in conjunction with the War and Navy Departments called "WAMS" (Women Apprentice Mechanics).[16] The program was developed to provide replacements for men called up for combat service. In 1943, 110 women were employed as "learners and helpers," but the estimates were that this could increase to the point where women filled almost 50 percent of the jobs. In general, 20 percent of TWA employees in 1942 were women, but by 1943 this grew to 35 percent.

There were 814 women mechanics working for Pan American Airways in December 1942. An article in the company magazine described the dramatic expansion of the feminine presence:

Numbers did not tell the whole story, by any means. Equally important was the excellence of the work being turned in by feminine hands and brains. Many jobs were, in fact, proving more suitable to female than to male skills, eg. cleaning spark plugs . . . .[17]

One of Pan Am's mechanic crews at La Guardia Field in New York was composed entirely of women. One was a former drugstore cashier, one was a sales girl, and three were housewives. In June 1943, this crew was considered by Pan Am to be the fastest and most capable on duty—male or "coed."[18]

Mechanic's Helper Betty Travis was the only woman in 1942 qualified to work directly on Clipper engines. Her previous experience in automobile engine repair indicates why she was able to assume this position so rapidly. Travis was clearly considered a model employee. The company reported that

four nights a week she is taking courses on motor transportation service, first aid and other subjects offered by the American Women's Voluntary Services, Inc. (AWVS), [and she] goes on duty as an air raid warden in her local district in Forest Hills one other night a week. She has also enrolled for the free courses in aviation mechanics soon to be sponsored in Pan American's La Guardia Field hanger by New York's Board of Education.[19]

The instructions to male employees at Pan American were clear—treat the women like men. False chivalry was not welcome, and for the first time there was a written policy on this matter. The most obvious manifestation of this order was that the men addressed the women only by their last names (the women spoke to the men in the same way). The company felt such changes in personnel policies (including the hiring of women) represented the "biggest opportunity air transport had offered women since the domestic airlines adopted flying stewardesses . . . ."[20]

The jobs for women in aviation were not limited to the air transportation industry. In fact, the vast majority of women were employed in some facet of aircraft manufacturing and production. Some of the positions traditionally held by women included work as public relations specialists. These women were expected to understand the industry thoroughly and to be able to answer both technical and non-technical questions about aviation. An understanding of research skills and journalism was considered an important requirement for this job. Frances de Marquis, for example, worked for Fairchild Aviation Corporation. Her work was principally concerned with preparing press releases on the aircraft built by Fairchild, but she also responded to the large number of individual requests for information about the company.[21]

The industry became eager in the early 1940s to train and employ women engineers, reflecting an acute shortage of suitably skilled applicants. A 1942 article

in *Independent Woman* by a woman engineer observed: "A year ago none of the plants wanted women. This year it's different and employment for women students who stick to their studies is virtually assured."[22]

Employment was "virtually assured" because in 1942 more than 18,000 new engineering jobs existed, but only 12,000 qualified students graduated to fill them. Women engineers in any aviation speciality were practically nonexistent in 1940. Few schools of engineering admitted women; the principal ones that did were Pennsylvania State University, Massachusetts Institute of Technology, University of Tennessee, University of Colorado, Cornell University, and the Newark School of Engineering. In 1942 two of the three women members in the American Institute of Electrical Engineers were employed in aviation—Edith Clark at General Electric and Mabel Rockwell at Lockheed. The ratio of female to male engineers was about one to 2,000 in the period 1940 to 1942.[23]

It occurred to the manufacturers, however, that they could use women with a mathematics or physical science background if they were given an intensive "crash" course in engineering. The Chance Vought Aircraft Division of United Aircraft Corporation, for example, established scholarships for women in aeronautical engineering at New York University. The program recruited women who had completed their junior year in college for an eight-month program of specialized technical training at the Daniel Guggenheim School of Aeronautics of the College of Engineering at New York University. The women, about 40 in each class, would receive their degrees from their original college through a transfer of credits from New York University. In order to receive the scholarship, the young women signed a formal contract with Chance Vought that included an option by the company on the student's services for one year after completing the training course.[24]

Rose Marie "Bonnie" Campbell (now Bender) of the second class of Chance Vought Scholars described the rigorous program. The first semester included mechanical drawing, calculus, and introduction to the three principal engineering fields: aeronautics, mechanics, and materials. In the second semester the women were taught aircraft design and drawing, aerodynamics, stress and weight analysis methods, and a survey of aircraft equipment and components. Those who successfully completed the course were taken on by Chance Vought in the Engineering Department. Their first six weeks were spent on rotation through each of the company's shops before receiving a permanent assignment. Bonnie Campbell, for example, worked on the stabilizer for the famous F4U-1 Corsair.[25]

Curtiss-Wright Corporation started an educational training program that was known as Curtiss-Wright Cadettes. Like Chance Vought, Curtiss-Wright also supplied tuition, living expenses, and a stipend for the women as they learned the profession. Curtiss-Wright, however, was much more ambitious. Its program involved 800 college sophomore-to-senior women who were sent to one of eight universities: Cornell, Northwestern, Purdue, Iowa State, Minnesota, Texas, Pennsylvania State, and Rensselaer Polytechnic Institute. Half were trained in the design of airframes, half in the design of airplane engines.[26]

Professors who had never taught women students before were surprised. At Rensselaer the faculty responded that the "Cadettes catch on in a hurry, ask more questions than do the boys, take the detail better, and therefore, learn their subjects more thoroughly."[27] After a 10-month course, the women were expected to work as professional engineers for Curtiss-Wright. These women were pioneers, seizing new occupational opportunities. They had a different status from that accorded to the trainees at Chance Vought and several other manufacturers that tended to see women as engineering aides and assistants.[28]

The aircraft industry was getting a bargain in its female engineering employees. The new sub-specialties they filled made it possible to get by with fewer "full-fledged" engineers, and the overall production process was enhanced. The new aides or paraprofessionals performed routine engineering tasks, such as calculations, drafting, and illustration, that required less education and training.[29]

A few women in aeronautical engineering were prominently featured in the news media. Elsa Gardner was one of them. In 1942 she was the only female aeronautical engineer in the Navy. Gardner had previously worked for several companies, including Eclipse Aviation and the Wright Aeronautical Corporation. She had also spent five years with the Army Air Forces as the editor of *Technical Digest* translating and condensing thousands of scientific articles on aviation topics from French, German, Italian, and English sources.[30]

Isabel Ebel was another engineering pioneer. She graduated from MIT in 1932, the only woman studying aeronautical engineering among a student body of 30 women and 3,000 men. Unable to find work in any American aircraft factory, she went back to school at the Guggenheim School of Aeronautics at NYU. No woman had ever been admitted previously, and only after the intervention of Amelia Earhart was Ebel accepted. Even with this additional academic work (she graduated in 1934), she was unable to find a company willing to hire her. Finally in 1939 she got her first opportunity, at Grumman Aircraft Corporation,

where for two years she worked on several airplanes, concentrating primarily on the XF5F-1. She then worked for two smaller firms until 1942, when she took a position as a research engineer with United Airlines.[31]

Isabel Ebel summed up the reception given to these remarkable female aeronautical engineers: "The fact that I am a woman has never hindered me with any engineering work I have done, but I don't know that it has given me any particular advantage either. . . . I have found on the whole that once the original barrier is down, women are fairly well accepted."[32]

There were in the war years more women in engineering than ever before, yet, in both comparative and absolute terms, their numbers were very small. General Electric, for example, employed only 12 women as professional engineers and 206 women in paraprofessional jobs such as computation or drafting.[33]

Although the female contributions to the war effort were generally recognized and appreciated by politicians, the public, and the press, it was commonly thought that the only possible motivation these women might have was patriotism and the desire to do something on behalf of a brother, father, or fiance. That opinion was, in fact, largely justified. Many women engineers sought training and employment for exactly those reasons. Bonnie Campbell, the Chance Vought engineer, wanted to use her mathematical knowledge to help the war effort. She announced her engagement shortly after graduation from NYU, however, and left the company right after her year's obligation was complete in order to get married.[34] Both the women engineers and the corporations perceived female employees as temporary additions to the factory work force. The experience of these women was therefore quite different from those who viewed themselves as a permanent part of the corporate world.

Three women who began work with the aircraft industry during the war and who did anticipate careers with their company were Cecil "Teddy" Kenyon, Barbara "Kibby" Jayne, and Elizabeth Hooker. They were hired to work ultimately as test pilots by Brewster A. "Bud" Gillies, a vice-president of Grumman Aircraft Corporation (and husband of Ninety-Nines president and WAFS second-in-command, Betty Gillies).[35]

He had become very concerned about the shortage of company test pilots, individuals who were critical to the manufacturing process. They flight tested every airplane as it came off the assembly line. The first to fly the aircraft, these pilots were exposed to a fair degree of risk in order to certify the aircraft for delivery to the military. Gillies believed that hiring women pilots was a very good solution to the problem. He was shrewd in making his pitch to the company. The three women he selected were active professional pilots before the war, and they were well educated and very confident in their skills. Gillies allowed them to demonstrate their competence in a non-threatening way to both the corporate officials and the male test pilots by first employing them as general pilots.[36] Starting in late 1942, Kenyon, Jayne, and Hooker flew short ferry runs to pick up, transport, and drop off materials for Grumman. Then they learned to fly the amphibious twin-engined Widgeons as well as the twin-engined JRF. At that point, it seemed that any opportunity to fly the company's fighter aircraft was just a dream for these women.

As soon as the three women had accumulated some experience with the company procedures, Gillies proposed to Grumman that they be allowed to work as production test pilots. His plan was a success. The company supported Gillies and had the women serve a trial period as test pilots on Grumman Hellcat fighter airplanes.

That decision had important repercussions. It reinforced the view that women pilots could operate combat aircraft safely and successfully. Further, it showed that women were as capable as men in the rigorous responsibilities of test piloting. Given the high media profile and visibility these women received, the three became vital role models to be cited by any advocate of women pilots. The three women's success was considered influential also in the decision to allow

FIGURE 5.—Mildred Strelitz was the first woman engineering aide to participate in test flights at Wright Army Air Field, Ohio, 9 November 1943. (Courtesy of D.G. Douglas)

women to become air traffic controllers.

There was tension at Grumman despite the good work of Kenyon, Jayne, and Hooker. Margot Roberts described this problem in her 1944 article about the three in *Woman's Home Companion*:

As a class, men fliers are nobly [sic] conservative, masculine and inclined to think of a woman's place as being in the home. They rarely welcomed lady pilots in their midst. This is something that all the gal pioneers of flying have bumped into at times. As they have the normal feminine desire to get along with men, a problem is created at first—though a very short-lived one in most cases—and you can't always blame the men. There are several angles to this man-woman equality proposition. At Grumman, for example, there is the matter of sticky ships.[37]

"Sticky ships" were airplanes that were suspected of having serious problems that could not be found by mechanics on the ground. The women were not permitted to fly these aircraft, nor were they permitted to do experimental testing such as dives. The three were annoyed by the arbitrariness of the policy, which also complicated their acceptance by male colleagues. The men were ambivalent, resenting the unfair exclusion of the women, yet also still none too sure about the notion of women flying at all. Further, even other women fliers, such as the prominent pilot and director of the WASP, Jacqueline Cochran, were critical of the job the Grumman women were doing. They viewed it as a kind of "aerial dishwashing," yet within the restrictions imposed on them, Kenyon, Jayne, and Hooker had ably demonstrated the competence of female test pilots. As Margot Roberts put it, "It *isn't* real dishwashing and [that is] one reason for the attraction."[38]

The vast majority of women in the aviation industry were production-line workers. The employment of women in such jobs predated Pearl Harbor and the American declaration of war, after which the number of women involved increased enormously with the dramatic expansion of aircraft manufacturing.

Beginning in 1938 with Hitler's annexation of Austria, then the occupation of the Sudetenland, followed by the invasion of Poland, events in Europe had galvanized the attention of Americans. On Sunday, 3 September 1939, Great Britain, and then France, declared war on Germany. In turn, the slow and inexorable process of American mobilization for an oncoming war began to quicken.

The first major manifestation of change was in the aircraft industry. Demand increased radically as President Franklin D. Roosevelt and his administration became explicitly committed to the premise that air power would win the war. The effect of this on the military was the abolition of the Army's ban on the development and production of bomber aircraft such as the B-17 and the B-24. Military leaders in the Army

Air Corps held a conference with the major aircraft manufacturers in order to establish plans for wartime production. That meeting was held in July 1939 and subsequently the manufacturers began to implement production of the two airplanes. The demand for aircraft already existed, and it was obvious that there would be a huge export market as the British and French became major consumers of American aircraft.[39]

Both the British and the French had identified the value of air power for offensive and defensive military programs. The disastrous experience with the German Luftwaffe in the Battle of Britain in the spring of 1940 reinforced the argument for bomber aircraft. Following the fall of France after the Dunkirk evacuation in June 1940, British Prime Minister Winston Churchill made an impassioned plea to the free world for assistance. It provoked an immediate response from President Roosevelt. Roosevelt persuaded Congress to provide the resources (political, economic, and military) in such measure that American industry would be capable of producing at least 50,000 airplanes a year.[40] Then, in January 1941, he worked to secure passage of the Lend Lease Act, which would extend the military's ability to lend, lease, or transfer American military equipment to foreign countries whose defense was considered vital to the United States. There was one other Congressional action stimulated by Roosevelt that would affect women in aviation. It was the establishment of the draft on 16 September 1940.

These events meant that the aircraft industry was faced with the need for a phenomenal growth of volume output, which would be hampered by a serious labor shortage. The industry catapulted from 44th (ranked by value of its output) in the United States in 1939 to first in 1944. Within the course of one year, from January 1942 to January 1943, total employment in all of the airframe, engine, and propeller plants of prime contractors changed from 460,356 to 1,027,914. These numbers do not include subcontractors and parts suppliers. There was a desperate need for workers, and as a result of the draft, women became the single greatest labor source for the industry. In that same 1942 to 1943 period, the total number of women employed by prime contractors grew nearly 1,300 percent from 23,137 to 321,788. Women represented 5 percent of the total general work force in January 1942 and 31.3 percent in January 1943.[41] In a wartime publication, *Education for Victory*, the lead article in June 1943 was on women in industry:

That old sign—"Men Only" no longer frowns unchallenged at the door of American Aviation. In many important spots of industry the ban on women workers has gone the way of the dodo bird. In others, it lingers feebly and outmoded, as untimely as last year's

FIGURE 6.—This parade float sponsored by Ryan Aircraft in San Diego, California, was designed to encourage women to join the aircraft production line.

circus poster. And in numerous shop windows appears a new sign—fresh, realistic and inviting—with a new slogan— "Women are Welcome."[42]

In June 1943 women represented 35 percent of the total number of workers, a proportion that rose eventually as high as 36.9 percent (July 1944). At the peak of employment in aviation (November 1943) there were 486,073 women (36.7 percent) among 1,326,345 workers.[43]

The transition was not made quite as easily, however, as the statistics might imply. Manufacturers were initially hesitant to use women, partly because of certain stereotypes regarding the kinds of work a woman could or should do, but mostly because of lack of experience with women in the workplace. High production demands temporarily suspended "philosophical" objections—women *had* to participate in the production lines if quotas were to be met.

It is noteworthy that much of this discussion refers to the experience of white women. Women of color found that they were faced with an almost impenetrable barrier of racism and sexism. They were often have proved, on the whole, good workmen; though the same differences in skills and temperaments may be observed among them as among men . . . ."[45]

At the individual supervisory level, the comments of Ryan Aircraft Factory foreman Frank Walsh express a typical reaction towards women workers:

I was dubious at first. Many of the jobs in this department required machine shop experience. But most of these girls had never even seen a machine shop. Yet today some of them are operating mandrels, burring and cutting machines. Furthermore, they occasionally turn out records that beat all our previous standards.[46]

Despite such positive comments about the new women, however, many negative ones were also expressed. In the same Ryan factory where Frank Walsh worked, one official stated:

Hundreds of the girls we have are not yet disciplined to stand the noises and smells of a factory. They often don't have the right temperament for this sort of work. In some departments they don't work as fast nor as accurately as men. And some of our girls are serious offenders in absenteeism.[47]

There is no historical evidence that women were worse "offenders in absenteeism" than men, yet this was something that became identified as a feminine problem.[48] Leslie Neville, for example, wrote in his March 1943 *Aviation* editorial that while

invasion of the aviation manufacturing industry by members of the weaker sex has done much to solve the problem of meeting labor force requirements of our greatly expanded production program . . . it poses new and far reaching problems for management and government . . . . the most serious—and still unresolved—difficulty that has come about since the ladies took to aircraft building is absenteeism.

Almost parenthetically, Neville added that this was not confined to women workers alone. The thrust of the article was that the workers, especially the women, were not fully aware of the serious nature of their work, that for some, working all-out for patriotic purposes was a novelty that "ebbs and flows with the headlines in their newspapers." Neville completely ignored the causes of absenteeism and argued that it was materialism that had consumed the new workers; in other words, the preoccupation of the workers was not "victory" but goods.[49]

Nothing could have been further from the truth. To participate in the aircraft factory life, women had to overcome several obstacles. First and foremost, they were not wanted. The most difficult prejudice for them to overcome was not the fact that feminine mechanical aptitude lacked social acceptance but rather the notion that women belonged in the home. The 15 April 1942 edition of the *Civil Aeronautics Journal* reported that "formerly custom, habit, the attitude of unions, the

FIGURE 7.—An unusual photograph of a black woman working on the aircraft assembly line, Eastern Aircraft Division of General Motors.

referred to the armed services if they wanted to work. The opposition to their presence in the aircraft industry would seem irrational since there was a greater cultural acceptance of a minority woman who worked to support herself and her family. There is also evidence that black women responded strongly to the call for female labor, yet examination of countless wartime photographs reveals few black "Rosie-the-Riveters" in the aircraft factories.[44] Integration of women of color into the aviation industry, as in many other industries, was a slowly evolving phenomenon.

Women were brought into three main areas as the factories tooled up for mass production. Most worked directly on the assembly line. A smaller group worked as clerks (both secretarial and production line). Slightly fewer were tool designers and makers. Male colleagues were gradually forced to concede that the women could do the job. Fairchild Aircraft Company issued a statement acknowledging that "they [women] attitude of management, and the considerations of cost, all served to curtail the number and kind of openings available to women."[50]

When Elinor Collins became a mechanic's helper in the fall of 1942 in Pan American's Alaskan Carpentry Section, it was reported that

eyebrows raised, the men started pools on which thumb she would blacken first. After a week, neophyte Collins reported to Chief of the Carpenter Shop, C.W. Smith without even a chipped fingernail. Shortly she was hammering nails, finishing wood and operating a table saw with the smooth sureness of a masculine master craftsman. "Yes, she's darn good," admitted fellow workers. "Mrs. Collins saws lumber the way my wife bakes a cake. She's brisk and efficient, never wastes a moment."[51]

The idea of a woman assembly line worker still grated on the American sensibilities, yet many women did manage to justify such activity in the name of patriotism and the war effort. The complexity of the definition and experience of patriotism made it a motivational tool of uncertain integrity. Propagandistic appeals to patrotism were frequently made in order to manipulate an individual, usually with the intent of producing some behavioral alteration based on a sense of guilt.[52] This tendency toward exploitation was recognized by several leaders, in particular by some connected with the employment of women in industry. Elinor Herrick, Director of Personnel and Labor Relations for the Todd Shipyards Corporation, wrote in the introduction to Laura Nelson Baker's, *Wanted: Women in War Industry*:

But such an appeal to this patriotism must not be the preliminary to their exploitation. They must have decent working conditions, they must have safeguards for health, they must be given full recognition of their efforts in terms of proper pay, promotional opportunity and the rest.[53]

The huge influx of new workers meant a heightened awareness of the conditions of the factory workplace; the presence of women insured media and public attention. Many managers such as the Ryan factory official quoted earlier implied that difficulties in adapting to the plant were a fault of gender.

The problem was not one of biology. Concurrently with the arrival of women—and other new workers—new labor laws developed. The Women's Bureau of the U.S. Department of Labor called for the promulgation of 13 new labor laws, which applied to males and females alike. These included the 8-hour day, 48-hour week, specified rest periods, a 30-minute meal period, minimum wage guarantees, extension of workmen's compensation legislation, and compulsory health insurance.

The laws were progressive (and expensive for industry), and were designed to "extend our legal bulwarks to guarantee on all labor fronts protection against the foes of fatigue, illness, accidents—against the hazards of unemployment and old age."[54] These proposed regulations were not uniformly adopted, but their influence was felt. Foreman Gordon Shop at the Consolidated Vultee Plant in San Diego commented that women had made a difference in the plant. He was quick to point out that women themselves had not made the changes; it was the company that instituted new time- and strength-saving devices. However, it is equally obvious that without the presence of women, these changes—procedures and devices that benefited everyone—would not have been established as soon as they were (if at all).[55]

It is questionable whether the production line problems of the expanded aviation industry were actually the fault of women's gender. There was little contemporary recognition that the difficulties might be explained as a function of inexperience, different socialization, education, and background. Individual women were often praised in glowing terms. Foremen and corporations exhibited considerable pride in their "girls'" accomplishments.

Women collectively, however, were often blamed for larger problems experienced in the factories at this time. Perhaps this was part of the reason why women, in far greater numbers than men, did not choose to continue working at the plants at war's end.

All kinds of women worked on production lines, either full time or as occasional help. During summer vacation periods, for example, school teachers would come to work with the aircraft companies. Two California high school teachers, Constance Bowman and Clara Marie (C.M.) Allen, worked the swing shift[56] at the Consolidated Aircraft Company building B-24 bombers. In the course of two months a significant transition was made. Bowman's book, *Slacks and Callouses*, reveals the moment that they first recognized the true value of their labor and their ability to make a substantive contribution. This occurred when they began to train their replacements, realizing that they were not "extras." "Our jobs were not going to be absorbed by other people in addition to their regular work. We had to be replaced, woman for woman, job for job."[57] This generated a feeling of pride that neither woman, despite her teaching career, had experienced prior to that point.

"You know, we really did do something this summer," said C.M. "Don't I know we did," I said looking at the ships and then my hands, with callouses on the palms and cuts on the knuckles and not one fingernail that extended beyond the tip of the finger.[58]

Bowman and Allen were not alone in feeling pride in their accomplishments. A talented aviation writer, Georgette Chapelle wrote two books (under the pseudonym of Dickey Meyer) during the war to encourage

FIGURE 8.—To help women adjust to factory life, many companies circulated posters with tips on how to manage the conflicting demands of work and home.

women to participate in aviation.[59] The key message of Chapelle's books and others like them was that women, while feeling proud about contributing to their country's war effort, also really enjoyed working in aviation. The latter feeling was what these authors believed would insure a continued feminine presence in the industry. For example, Chapelle wrote:

The aircraftswomen of today are proud of their skills, but they know that the system by which they work permits no prima donnas. The girls remember well the days when femininity and aviation met only in headlines and press pictures, and they do not propose to revive that unfortunate custom.

They have begun a new tradition, and one which they believe will outlive the other by many generations. It is a tradition of accomplishment, of consistently getting things done. They are raising their standards ever higher each working day, and beneath their uniforms and coveralls their hearts have wings.[60]

What is especially notable about the period of World War II is that in the aircraft industry, women were for the first time permitted to enter the higher-paying manufacturing jobs instead of being relegated exclusively to lower-paying non-manufacturing jobs (usually secretarial work). Yet even as these changes were occurring, it is apparent that women were deliberately placed in specific occupations.[61] Starting out with skills equivalent to those of the men, women still found they were channeled into the dullest, most tedious occupations—all the while being praised for their wonderful ability to do inspection, sewing, small parts assembly, and rivet checking. A factory supervisor commented:

Men get tired of doing the same thing over and over again but some of the women will stick right by the job hour after hour. Jobs often have to be broken down and simplified for them, and extra-fine tooling is required—but once a simple job is set up for them they'll go to town on it without ever showing any signs of boredom.[62]

Women were assigned to jobs requiring fine, precise, detailed work such as inspection. Yet this work was considered to be of a lesser status in the factory than those jobs more commonly filled by men. Most factories rated their employees according to a specific code; the higher the rating, the more valuable the employee. Supposedly this was because the highest rating meant the greatest degree of versatility. In fact, in most companies it was usually "physical strength," a quality that was neither an asset nor a requirement for many factory jobs, which received the priority code rating, and consequently the highest status.

Men still retained control of almost all the supervisory positions. Even when women worked in similar jobs they often deferred to the men. Josephine R. Viall was the wife of a Marine Corps major as well as an editor of a Marine Corps magazine. When her husband

was moved to the Pacific battle zone she started working at the Ryan aircraft factory. She was bright, able, and capable of leadership. Yet, when she was placed in a new environment she attributed her successful orientation exclusively to the efforts made by others (in this case, men). Her comment was typical of the expressions made by many women: "The men in the factory have been wonderfully patient in teaching me my job—even showing me around the factory so I'd see how my little operation fits in with the whole big picture."[63]

In terms of the big picture, most women on the production line were in "dead-end" jobs, and on average they were paid less. The average hourly earning of women in aircraft engine plants in August 1943 was $1.04, or 87 percent of the men's $1.19. In aircraft metal propeller plants the average hourly earnings for first shift workers in October 1942 were $.81 for women and $1.05 for men.[64] All women faced the possibility of eventually being fired because the men were guaranteed their jobs after the end of the war.

A gradual shift in war propaganda began to occur in 1944. A hunger for "the good old days" had set in. Advertisements, which had at one time encouraged women to work in factories, began to promote their eventual return to the home and domesticity. Ambition was not encouraged. In aviation the situation was somewhat different. Many thought (naively) that war-level production rates would be sustained, even expanded, after the hostilities ceased. These individuals believed, therefore, that women would still be needed and desired for aviation work. Even among those who were not caught up in the visionary rhetoric of an impending "aerial age," there were many who argued that hardworking, capable women would not be fired.[65] Although the view was a mistaken one, progress had been made and recognized.

The war is teaching that many women have exceptional mechanical aptitude and can operate ponderous and complex machinery, sometimes more skillfully than men . . . . War teaches that nearly every man and woman can be used. And we are using many thousands of them. Many sociological thinkers believe that the social effects of these lessons that we are learning today may be more lasting than victory.[66]

"Rosie-the-Riveter," women mechanics, and their white-collar companions in air transportation and aviation manufacturing had a profound effect on American social history. The labors of these women were not part of any planned policy, such as the affirmative action programs of the 1970s and 1980s. Wartime mobilization had simply required a huge work force. Most women entered the work force through programs that were considered experimental and

FIGURE 9 (above left).—Four women work on sewing the fabric cover for an Avenger torpedo bomber rudder.

FIGURE 10 (above).—Sorting rivets by size was the job of many female aircraft workers.

FIGURE 11 (left).—A woman carefully orders the pattern of wires and cable forms that will comprise the electrical systems of either a Wildcat fighter airplane or an Avenger torpedo bomber airplane.

temporary. The voluntary participation of all citizens in the war effort was the goal of such programs, which is why working women in aviation became identified primarily as "volunteers" instead of as permanent workers or professionals.[67] Despite these circumstances, women in the labor force exerted considerable influence on postwar America by removing many of the barriers against married working women.

White women found there was a clear distinction between occupations that accepted (and often assisted) married women (such as work in aircraft factories) and those that did not (such as flight attendants). Black, Hispanic, and other minority women had an extremely difficult time trying to obtain war work, and for most Asian-American women the situation was impossible.

The military was slightly more tolerant. Almost every civilian occupation, except aircraft assembly, had a direct counterpart in the military. The story of military women in aviation during the war will be explained in the next two chapters and is critical to understanding the total wartime experience of women in aviation. However, by virtue of the sheer numbers of women in the aviation industry, it is the image of "Rosie-the-Riveter" that wrought the most change in American attitudes towards working women in the post-World War II age.

3. Daughters of Minerva: Military Women in Aviation

For this is what the WACs declare
To lads the world around:
"You man the ships and guard the air
And we will guard the ground."

from "Air WACs," author unknown

The development of air power in World War II encouraged many women in the belief that they could contribute to the defense of the United States. Female pilots were as eager as the men to put their flying talents into service. These women were not content to be shunted into peripheral roles; they wanted to be an integral part of the military. They wanted to be "daughters of Minerva," the Roman Goddess of War.

The idea of women in the military was actually not a novel one. American women had been serving the armed forces since the Revolutionary War, although often their service was either clandestine or auxiliary in nature. There are narrative accounts and documentary evidence dating back to the nation's first military struggles that confirm the service of women in combat, despite regulations officially prohibiting their recruitment. Little is known of these women because of the tendency of nineteenth and early twentieth century historians to assume that they must have been servants, camp followers, or prostitutes and therefore unworthy of mention in historical texts.[1]

Despite this meager historical record, it is a documented fact that women were asked to serve as enlisted personnel in World War I. The Navy, finding no explicit bar based on gender, recruited women as "Yeoman (F)" for the Naval Reserve. More than 11,000 Yeomen (F) were in service (primarily as clerical workers) on Armistice Day, 11 November 1918.[2] The experience of that female Navy contingent was largely forgotten or ignored by later military planners and the public, although some of the leadership and many of the ideas for the various women's corps proposed during World War II came from these Navy women. But during the interwar years, the notion of women on active military duty was at least open to debate. Indeed, the degree of approval and acceptance the idea gained in the 1920s and 1930s was crucial in determining the extent of female participation in World War II.

Women had been exploring the potential avenues of military participation for some time. The well-known aviator Louise Thaden devoted an entire chapter in her 1938 autobiography, *High, Wide and Frightened*, to a fictitious story about two women flying combat missions. She wrote of women brought into the Army because of a severe shortage of trained, experienced male pilots. The story's two young protagonists are gradually drawn into increasingly dangerous assignments, all the time worrying about personal failure lest it result in all women pilots being grounded. In the end, one of the women is severely wounded, losing a leg, but despite this personal tragedy, the message conveyed is that one individual's injury should not affect women's participation in general, and that injury in the line of duty should be seen as a socially accepted hazard, regardless of sex.[3]

The use of women in aerial combat was not seriously considered before the United States entered the war. There were, however, serious proposals for employing women flyers in the military. It is hard to find accurate records of the many plans that emerged in the early days of World War II. There was considerable overlap and some duplication of programs being developed simultaneously. Of greatest importance, the *idea* of women in military service received wide acceptance among individuals who, by virtue of their positions, could influence policy and effect change.[4]

In September 1939, the same month in which she had set a new women's international speed record,[5] Jacqueline Cochran wrote to Eleanor Roosevelt about the use of women pilots in a national emergency. The letter would have important repercussions. Mrs. Roosevelt was very interested in women in aviation. Both she and the President had been good friends of Amelia Earhart, and Mrs. Roosevelt's personal desire to learn to fly was frustrated only by the Secret Service, which feared sabotage. Mrs. Roosevelt had written about aviation in her famous "My Day" columns, and later, during the war, she became a constant advocate for women pilots. Mrs. Roosevelt also knew of Jacqueline Cochran's background.[6]

Cochran was an indefatigable (and sometimes irascible) woman. Born into extreme poverty in northern Florida, Cochran's later triumphs remind one of a combination of Cinderella and Horatio Alger's "Raggedy Dick." In her autobiography, Cochran describes her life in these terms: "It is a story of flights and fancies, of privations and places and perfumes and laces and of aces and kings and generals—all scrambled together."[7]

"Perfumes and laces" symbolize the first of her significant experiences, for Cochran began working in beauty shops at about the age of 11. This was a violation of child labor laws, but her work was essential in helping to support her family. Soon after this, she left home and started working in another shop where she eventually became a hairdresser. In the early 1930s she began working for Antoine's, a beauty salon for the extremely wealthy. Later Cochran, still a hairdresser, worked at both the New York and Miami Beach locations of Saks' Fifth Avenue, following the seasonal migration of her rich patrons.

Often invited to parties by her clients, she chanced to meet millionaire Floyd Odlum, a Wall Street financier who ran Atlas Utilities and Investors Company, Ltd. Cochran described to him her dream of starting her own cosmetics company. Her plan was so elaborate that Odlum suggested that in order to succeed she would need wings. Odlum's casual remark triggered Cochran's imagination, and in 1932 she spent a three-week vacation learning to fly at Roosevelt Field, Long Island.[8]

Cochran's fledgling cosmetics business proved a success, and her flying exploits were soon exceptional. Having entered her first major air race in 1934, she was soon breaking air record after air record. Then in 1938, she won the prestigious Bendix Trophy Race in a Seversky pursuit plane.[9] She was the Harmon Trophy recipient in 1938 and in 1939 she was awarded the Harmon Trophy again as well as the General "Billy" Mitchell Award, which was given to the "American pilot who during the previous year had made the greatest contribution to aviation."

In 1936 Cochran had married Floyd Odlum. He was a friend of the Roosevelts (and many other prominent politicians) and a major contributor to Franklin Roosevelt's political campaigns. So it was that in September 1939 Cochran's letter to Mrs. Roosevelt was assured a warm reception.

Cochran wrote to the First Lady about using women pilots, asserting that "in the field of aviation, the real 'bottle neck' in the long run is likely to be trained pilots. Male pilots could be released for combat duty by assigning women to all sorts of helpful backlines work."[10] Later during the fall, she made her pitch to the Ninety-Nines, but it was not until March 1941 that her ideas began to get serious attention. Cochran had served on the Collier Trophy Committee that year, and after the presentation made by President Roosevelt at the White House she went to lunch with General Henry H. "Hap" Arnold and Clayton Knight, the acting head of an American recruiting committee for the British Ferry Command. During lunch, Great Britain's dire need for pilots was discussed. Cochran was asked to help recruit pilots, but a greater opportunity soon presented itself when General Arnold suggested that she actually do some of the flying. This, he thought, would be the best demonstration of Britain's genuine desperation.[11]

There is evidence that General Arnold and members of his staff were interested in the use of women pilots even before this encounter. Early in 1940, Lt. Colonel Robert Olds, then in the Plans Division of the Air Staff, had requested Nancy Harkness Love to list all the women pilots in the United States holding commercial ratings. Love complied, drawing from the Aero Chamber of Commerce lists. She also listed the women who had particularly outstanding records. There were 49 names on this latter list. Love requested that Olds take her name off, modestly stating she was only "obeying orders" by including it in the first place.

Nancy Harkness Love well deserved her place on that list. Having learned to fly in Houghton, Michigan, in 1930 at age 16, she exhibited a lifelong passion for aviation. She was well educated (Milton Academy and Vassar College), and her parents supported her varied interests. Her father (if not her mother) was enthusiastic about her learning to fly. She earned her commercial license while in college. Although she was forced to leave Vassar after her sophomore year in 1933 because of the Depression's effect on her family's finances, she continued to fly, and in 1935 she was one of three women hired by the Bureau of Air Commerce to work on its air-marking project.[12]

Married to Robert Love in 1936, she discovered on her West Coast honeymoon (flying, of course) that the Beechcraft Company had entered her in the Amelia Earhart Trophy Race at the National Air Races in Los Angeles. With no experience at pylon flying, she attempted to back out. Unsuccessful in her attempts to avoid the race, she managed to finish in fifth place.[13]

Back in Boston, she worked for the Gwinn Aircar Company, a job that included flight testing a new tricycle landing gear. She and her husband had just started what was proving to be a very successful aircraft sales company called Inter-city Airlines, when the war in Europe began to have an impact on her life.

As a well-known pilot, Nancy Love, like Jacqueline

Cochran, participated in ferry flights. In June 1940 she and other United States pilots were responsible for flying American airplanes to the Canadian border, literally pushing them across the international line, and then flying them to their Canadian destination, where the planes would await shipment to France. The airplanes were to have been used by France, which was under siege by the German Luftwaffe. Shortly thereafter, however, the German Army occupied France.[14]

The flights brought Nancy Love into contact with the operations of the Army Air Corps' Air Ferrying Command (which was known after 9 March 1942 as the Ferrying Division of the Air Transport Command). Her connection with this organization was later reinforced by her husband's duties. Robert Love was a reserve officer in the Army Air Corps and, in the spring of 1942, he was recalled to duty in Washington as the Deputy Chief of Staff of the ATC.

Robert Love's new job also brought his wife to Washington. She obtained a civilian post with the ATC Ferrying Division operations office in Baltimore, Maryland, commuting daily 80 miles roundtrip by airplane. In the interim between 1940 and 1942, possibly at Colonel Olds' suggestion, Nancy Love had continued expanding on plans she had in mind for using highly qualified women pilots in a military effort. It is not clear, however, to what extent Love had promoted such a plan before her move to Washington. Because women had continued flying actively with the Civilian Pilot Training Program as instructors, even after they were banned as students in June 1941, the promotion of a military women's flying unit may not have seemed terribly urgent prior to early 1942.

American military leaders had turned their attention to the impending problems of full scale mobilization as war clouds became ever more threatening over Europe in the late 1930s. Within this context there was planning for a women's army corps as early as October 1939. A memorandum that was prepared in conjunction with this effort recommended the use of women by the Army but only in a quasi-military unit. "Women under no circumstances [would] be given full military status."[15]

With the passage of the Selective Service Act in September 1940, women's groups began to apply pressure on Congress and military leaders for permission to contribute. In March 1941, as Congresswoman Edith Nourse Rogers was preparing to introduce legislation to provide for a women's corps, General George C. Marshall expressed the official military position toward women:

> While the United States is not faced with an acute shortage of manpower such as has forced England to make such an extensive use of women, it is realized that we must plan for every possible contingency, and certainly must provide some outlet for the patriotic desires of our women.[16]

The tone of Marshall's statement indicates a mellowing of official rhetoric, which had previously opposed any female participation, but it was obvious that until shortages of personnel became extreme, authorization would have a difficult passage through Congress.

The pivotal moment for women and military aviation occurred with the bombing of Pearl Harbor. The time lag between Pearl Harbor and the founding of the various wartime women's aviation groups resulted only from the need to pass enabling legislation and to set up the requisite organization structures. Military aviation jobs increased, in large part because of factory output: the sudden appearance of a multitude of new aircraft resulted in pilot shortages for both ferry and combat missions. Women in the aviation industry were already "in action" well before the military would actually enlist female personnel.

At the same time that Nancy Love was working on the earliest plans for the Women's Airforce Service Pilots, Jacqueline Cochran was in the process of arranging a ferry flight for herself to England. The British Air Transport Auxiliary (ATA) was not quite as enthusiastic about the participation of Cochran and other American pilots as Clayton Knight had been. In fact they dragged their feet until Cochran appealed to the newly appointed British Minister of Procurement, Lord Beaverbrook, William Maxwell Aitken, who was a good friend of both Cochran and her husband. In early June 1941, the ATA invited her to their Montreal base for tests and check flights. Cochran passed the tests (which were extremely stringent) and was assigned to ferry a Lockheed Hudson bomber to Prestwick, Scotland.

Mass protests arose from the male ATA pilots, who did not want to be blamed if the Germans shot her airplane down. Further, it seemed they felt that the presence of a woman trans-Atlantic pilot would demean their position. Pride, prestige, and entrenched prejudice were really the issues. A compromise was worked out between the ATA and its objecting pilots: Cochran would pilot the airplane in flight, but her copilot would be responsible for takeoff and landing.

After she arrived in the United Kingdom, Cochran met with one of England's premier women pilots, Pauline Gower, who was the chief of the women fliers with the ATA. Gower asked Cochran if it would be possible to recruit American women pilots to augment the British group. The request triggered a responsive chord in Cochran.[17]

Immediately upon her return to the United States in July 1941, Cochran was invited to lunch by President and Mrs. Roosevelt. She spent two hours discussing with them the current state of Britain and the war in Europe. In particular, President Roosevelt was very interested in her assessment of the Royal Air Force. A few days later, Mrs. Roosevelt asked to speak with Cochran again. This time she wanted to discuss the use of women pilots, especially in the United States. President Roosevelt then requested that she research this issue and, in support of this assignment, he sent a letter of authorization to Robert A. Lovett, Assistant Secretary of War for Air.

Lovett arranged for Cochran to be officially appointed (without pay) to work with General Arnold and General Olds (who had been promoted by this time). General Olds was very interested in hiring a select group of highly qualified women pilots to ease his immediate shortage in the Air Ferrying Command, a plan that had gradually emerged from his requests for information from Nancy Love more than a year before. Cochran worked furiously with CAA records to identify all the commercially rated women pilots. Then she sent questionnaires to them, and 130 of the 150 on the list responded. All who responded were enthusiastic about the possibility of flying for the military. Finally, Cochran matched the respondents and their skill levels with aircraft due to be delivered to the military, in order to show that not only were women pilots available, but also that they had the requisite skills to fly the military aircraft in question. The resulting proposal package was submitted on 30 July 1941.[18]

A prototype group of crack women pilots was to be employed on a trial basis. If successful, a full-scale women's pilot division would be organized and commissioned in the Air Corps Specialists Reserve. Cochran naturally assumed that she would be retained as chief of that division. General Olds and Cochran disagreed, however, on one major point of this proposal. Cochran wanted to create a women's organization that would include an on-going training program. Otherwise, she said, "the female effort would be a flash in the pan."[19]

General Olds was not willing to recommend the creation of a significant new military subgroup, especially one composed of women. Cochran was resolutely committed to her idea. Unable to get her way, she resigned from her position, but not before she presented her case once more to General Arnold. Arnold verbally assured Cochran that he agreed with her in principle, but he convinced her that the present moment was not right for her program. He suggested that she continue to elaborate her plan, in the light of anticipated pilot shortages.[20]

Much as he may have personally liked and respected Cochran, Arnold in fact actually disagreed with her proposal. Like General Marshall, he was unconvinced that the shortage of personnel required the establishment of all-women divisions. It was too experimental, too controversial, and not really necessary because there were more pilots than aircraft in the Air Corps at that point in the summer of 1941. The result was that Cochran returned to New York and began to recruit a team of women pilots to serve with Britain's ATA.

The men and women of the ATA were described as doing the "grease monkey work of the airfields." Shuttling back and forth among bases, factories, and foreign countries, they delivered all the airplanes used by the Royal Air Force (RAF). Women had been admitted in early 1940 but were initially permitted to fly only small trainer aircraft. They quickly proved that female pilots were both capable and competent. All restrictions were removed, and the ATA began actively to recruit pilots, both male and female, from other nations.[21]

Jacqueline Cochran hand picked a group of 24 women to form an American contingent (see Appendix I). She viewed this undertaking as an important means of demonstrating the viability of her proposals to the American military establishment as well as a way of keeping herself in the forefront of women in military aviation. The women who were chosen were not only highly skilled but also of upstanding character. Cochran did not want to risk failure because of personality problems or the appearance of less than exceptional moral behavior.

Cochran had selected her group by late 1941, and the women began training. They received check flights in Canada first, before going to England to begin conversion training. One American, Virginia Farr, a young flight instructor before being recruited by Cochran to join the ATA, described her work to a friend in the States: "This work with the ATA—flying anything and everything of two motors or less, and trying desperately to keep them all in one piece is a real experience...."[22] Farr continued with the ATA for three years even though she, like all the other American women, was only committed to 18 months of service. The group was recognized as outstanding. All were good pilots, all were dedicated to helping Great Britain. Several American women were injured, and one, Mary Nicholson, Cochran's personal secretary, was killed when the propeller came off the aircraft she was ferrying.[23]

Some of the British pilots resented Cochran's ostentatious style. She accepted the rank of honorary Flight Captain even though she did very little flying, and she usually appeared at the field in a mink and a

FIGURE 12.—Ann Wood was one of the 24 women picked by Jacqueline Cochran to fly with the Air Transport Auxiliary of Great Britain. (Courtesy of Ann Wood)

Rolls-Royce. She was, however, extremely effective in getting a job done, and in September 1942, she decided the time was right to make another pitch to General Arnold for an American women's air corps.

In the meantime, since Cochran's first proposal to Arnold in late July 1941, progress had been made in the creation of other womens' corps to serve with the various branches of the military. The women's air corps idea did not develop in a legislative vacuum. The WASP, WAC, and WAVES groups developed simultaneously, and advocates for including women in the military used progress in one branch to promote advances in the others. There were elements of bluff, shrewd manipulation, and honest lobbying that engaged virtually all offices within the executive and legislative branches of government.

The crucial starting point is 28 May 1941, when Congresswoman Edith Nourse Rogers introduced H.R. 4906 to establish a Women's Army Auxiliary Corps (WAAC), a group not *in* the Army but rather *with* the Army. Hoping that this distinction would make the

difference in gaining support, the bill's proponents carefully lobbied Congress. During the summer of 1941, even General Marshall became an enthusiastic supporter. He saw that women could be used effectively to counter future labor shortages; hence, he wanted in advance of any emergency to have the authority to recruit and train such a group. Still Congress balked, relegating the bill to the Bureau of the Budget. Here the measure sat until 11 December 1941 when all objections were withdrawn.[24]

Also, in December 1941, the Army Air Forces began to exhibit considerable interest in creating an Air Force section within the proposed WAAC. Correspondence between General Headquarters, Air Forces, and the Chief of the Air Corps [sic] went so far as to state: "This headquarters would recommend a separate and distinct organization, except for the fact that there should be only one women's corps serving with the army."[25] In January 1942, the War Department indicated that an aircraft warning section would be a part of the proposed WAAC.

Congress finally passed the Rogers bill on 15 May 1942 and the next day, Oveta Culp Hobby was sworn in as director of the new organization. From the time of the AAF Headquarters letter of December 1941 until the actual arrival of the first WAAC members on 20 September 1942, the Army Air Forces assumed the lead in developing plans for the use of female troops. The AAF leadership had a much broader view of employing women than retsricting them to the roles of typists or telephone operators.

Nancy Love's and Jacqueline Cochran's early efforts made a significant impact on the AAF views, even though their proposals for using women pilots had not yet been implemented. The establishment of the Air WAACs (later Air WACs) and that of the WAFS/WASP were closely intertwined, although the female directors of these programs refused to acknowledge this reality. Each organization set precedents for the other, and this entwined relationship must be understood in order to appreciate the subsequent events of these organizations' histories.

As early as November 1942 the suggestion was made that the Air WAACs be used as Link trainer instructors. The AAF was also very eager to have its women auxiliary members trained as aircraft mechanics and radio operators.[26] However, the first significant assignment of the Air WAACs was with the Aircraft Warning Service. The AWS was a vital defense component, constantly monitoring radios and other electronic devices for signs of enemy air attack. As this threat appeared to diminish, the AAF began, in early 1943, to phase out this use of the Air WAACs and to operate the stations exclusively with civilian staffs.[27]

FIGURE 13 (above).—Women's Army Corps officer is shown instructing a student in a Link trainer. (NASM/USAF neg. 27503A.C.)

FIGURE 14 (left).—Private Betty S. Dayton, WAC, works as a gunnery instructor for a U.S. Military Academy cadet. Pfc. Dayton flies the enemy airplane on a projection screen and instructs a cadet in shooting his target. (NASM/USAF collection)

FIGURE 15.—Colonel Oveta Culp Hobby, Director of the Women's Army Corps, talks to a group of officers before boarding an AAF aircraft. (NASM/USAF neg. 29772A.C.)

This development should hardly be perceived as the AAF's lack of interest in its female staff. On the contrary, the WAAC strength limit had been raised the previous fall from the initial 25,000 to 150,000, and the AAF's quota of the total WAAC strength was 65,000, or about 43 percent. Long range recommendations had been made to expand the Air WAACs to 1.5 million women by 1946.[28] This projection was based on the conclusions of a study by the Adjutant General's office that 65 percent, or 406, of the 628 military occupation specialties could be performed by women.[29]

The impetus for this study was to be found in the efforts to convert the women's auxiliary status into true military rank. The War Department decided it was desirable to integrate the WAAC into the Army to prevent further administrative and legal difficulties. Legislation was introduced in 1943 to set up the Women's Army Corps (WAC), and on 1 July 1943 the new group was officially established. All the women who had served as WAACs were offered positions in the WAC, but they had no legal obligation to enlist.

Regardless of the total strength goals, the War Department had, from the beginning of the WAAC program, made a firm commitment to ensuring the

FIGURE 16.—Sergeant Lois Herring, WAC, bench-tests an aircraft radio set, prior to its actual installation in an airplane. Sgt. Herring frequently went on cross-country flights as a radio operator. (NASM/USAF neg. A-33930A.C.)

that was where other black women were being assigned. In all cases, black women moved as a unit. Thus, black women were stationed at only 10 AAF bases in the United States, and only at Douglas Army Airfield were there black women working as aircraft mechanics or on the flight line. The system in general excluded black women because of prejudice, so recruiters were never successful in attaining the 10 percent goal. At best, the peak strength was about 4,000, representing four percent of the corps.[31]

FIGURE 17.—Corporal Evelyn Rivers, WAC, was a teletype operator at Lockbourne Air Base, Columbus, Ohio. (NASM/USAF neg. 34034A.C.)

presence of black women among the ranks. The department, with full support from Director Oveta Hobby, proposed that up to 10 percent of the WAAC personnel be black. (This carried over for the WACs as well.)[30] They did not expect the criticism that came from several national black leaders; for example, Dr. Mary McLeod Bethune protested the WAAC's plan to follow the Army policy of segregated troops. The idea of integrated troops was hotly debated but in the end, did not prevail.

Some black women did serve as Air WAACs (and later as Air WACs). In fact 80 blacks out of a total of 440 women were part of the first officer-candidate class. Assignment of black women was made quite difficult because of the segregated troop policy. This often meant that more talented individuals did not receive the assignments or opportunities they deserved because they were forced to go to the same location as their black colleagues. For example, musicians would be assigned to work as cooks or postal clerks because

FIGURE 18.—Private "Mike" Stanton, WAC, was the only 1st Air Force WAC qualified to "preflight" a single-engine airplane. Pfc. Stanton is shown servicing her Republic P-47 known as "Mike's Baby." (NASM/USAF collection)

FIGURE 19.—Private Doris Smith, WAC, worked on the flight line at Robins Field, Georgia. Here, Pfc. Smith is turning the propeller on a Stinson L-5 airplane. (NASM/USAF neg. 36038A.C.)

No other racial minority group enlisted in the corps in significant numbers either. Puerto Rican, Chinese, Japanese-American, native American, and other women of color did serve, and for the most part they were not segregated. However, they were recruited only rarely. Language often represented a problem and could prevent a woman from being assigned to a technical aviation specialty.[32]

At the same time the WAAC was gearing up, the Navy was also developing its own women's group. Congress-woman Edith Nourse Rogers' interest in a women's corps was not limited to the Army. In December 1941,

Rogers telephoned Rear Admiral Chester W. Nimitz to ask whether the Navy was interested in a bill similar to H.R. 4906 (the WAAC authorization introduced the previous May). Nimitz replied that clerical positions could potentially be held by women, but that she should ask the Secretary of the Navy for an official view. Rogers did this, thereby stimulating the Navy bureaucracy to recognize that the formation of some sort of women's organization was probably inevitable. In that event, they preferred to devise their own plan rather than cope with "unworkable" congressional legislation.[33]

"Indifference and scant imagination" were the major problems within the Navy. Only the Chief of Naval Operations and the Bureau of Aeronautics (BuAer) evidenced enthusiasm about the prospect of women serving. Of these two, it was the Bureau of Aeronautics that envisioned the use of women in a variety of technical and skilled positions. Consequently, the policy suggestions made by BuAer greatly influenced the major aspects of the organization and the structure of the eventual women's reserve section.[34]

Joy Bright Hancock had served as a Yeoman (F) in World War I. During that time she was the first woman to take the tests for Chief Yeoman, and despite opposition, she earned her rating. She had learned to fly in 1930. She also worked for many years as a civilian with the Bureau of Aeronautics, first at the Naval Station in Lakehurst, New Jersey, and later as head of the bureau's editorial and research section. While working in the latter position, Hancock had been given the responsibility in late December 1941 for outlining the ways that women could participate in the Bureau of Aeronautics. Based on these proposals as well as the introduction of the WAAC legislation the previous May, the Bureau of Personnel recommended on 2 January 1942 that the Secretary of the Navy ask Congress for authorization to create a women's section in the Navy.[35]

The Navy officially proposed in early February to add a title to the Naval Reserve Act of 1938. Title V, Section 501, proposed a unit that was called a women's auxiliary, but in fact the group was to be fully integrated into the Naval Reserve. The Bureau of the Budget rejected the proposal out of hand, suggesting that the Navy reformulate its plan to follow the WAAC legislation.

Meanwhile the Bureau of Aeronautics was becoming so eager for legislation to pass that it initiated a lobbying effort of its own. Using political connections in both the House and the Senate, Title V, Section 501, was introduced. It passed through the House Naval Affairs committee but met with opposition in the Senate. That opposition resulted in a revision that

FIGURE 20.—Air traffic controllers at Randolf Field, Texas: (left to right) Sergeant Jean Daubert, Corporal Rose Chytal, and Corporal Lois White. (NASM/USAF neg. A-26421A.C.)

compromised several unique features of the bill which would have permitted women to fly naval aircraft. Even with the proposed changes, the legislation was opposed by Senate Naval Affairs Committee Chairman David I. Walsh. Walsh viewed the entire idea as a threat to femininity and motherhood.

This viewpoint did not prevail, however, and the Bureau of Aeronautics' lobby came to an understanding with the committee that it would report favorably on the legislation once it was rewritten along the lines of the WAAC model. TheNavy (including BuAer) adapted the legislative proposal and grudgingly sent this new version to President Roosevelt on 25 May 1942. The President signaled his approval, mistakenly believing this to be the plan desired by the Secretary of the Navy.

In a last ditch effort to have the original plan (which proposed an integrated women's corps) approved instead, the Navy asked Roosevelt to reconsider. They outlined the problems with the Senate that had occasioned the compromise version, but it was not until the intervention of Eleanor Roosevelt that the President actually decided to investigate the differences between the two plans. About the same time the revised Navy legislation was sent to the President, Dean Harriet Elliot of the University of North Carolina had written to Mrs. Roosevelt, outlining the details and rationale of the original Navy plan. Mrs. Roosevelt shared that letter with the President. Roosevelt, fully comprehending the situation, promptly informed Frank Knox, the Secretary of the Navy, that the secretary had "*carte blanche* to go ahead and organize the Women's Reserve along the lines [he thought] best." The choice was made for a group fully integrated into the Naval Reserve. Congress acquiesed and the authorization bill, Public Law 689, was signed into law on 30 July 1942.[36]

The newly formed women's reserve program was called WAVES, which stood for "Women Appointed for Volunteer Emergency Service" (later "Appointed" was changed to "Accepted" because the former applied only to officers). From the start the Bureau of Aeronautics made them welcome. They asked for 20,000 WAVES immediately, a number that staggered the imagination of the planners at the Bureau of Personnel, which had estimated the total need at only 10,000 women (and those primarily in the clerical field).[37]

The Navy exhibited considerably greater foresight than the other branches of the service. Drawing on the academic world, the Navy deliberately requested the assistance of faculty and administrators from several of the prominent women's colleges as well as noted female professionals from coeducational institutions. The members of this advisory council had a great deal of practical experience in dealing with women in institutional settings. They were particularly effective in setting up the basic administrative structure as well as in selecting the WAVES' first director, the dynamic and erudite Mildred McAfee, president of Wellesley College.

Holding first the rank of lieutenant commander and later captain, McAfee was acknowledged as an outstanding administrator throughout her tenure as director. Enthusiastic, intelligent, and gracious, McAfee was able to get along with everyone, from "Old Salts" and naval brass to young women enlistees. She was very interested in promoting "women in naval aviation" and one of her first decisions was to appoint Joy Bright Hancock as the Women's Representative to the Chief of the Bureau of Aeronautics. This meant that Hancock would also be in charge of all the WAVES assigned to BuAer.[38]

The women who served as WAVES came from many places, but they were a homogeneous group. A study published in *Naval Aviation News* in June 1943 characterized the majority of enlistees as 22 years old; 5 feet, 5 inches tall; 124 pounds; with brown hair. They were white, single, and most were high school graduates employed as office workers prior to enlisting. Many were motivated by desire for a dramatic departure from their previous life, along with a patriotic enthusiasm for serving the war effort.[39]

WAVES in naval aviation served in many capacities. At the end of the war, the list of aviation officer billets (jobs are called billets in the Navy) for women included aerological engineering, aeronautical engineering, air-combat information, air-navigation gunnery instructor, air transportation, assembly and repair vocational training, celestial navigation (air navigation), flight desk, flight records, Link training, photographic interpretation, recognition, recognition and gunnery, radio-radar, schedules, and air traffic control. Enlisted women served as aerographer's mate, aviation machinist's mate (for both aircraft and instruments), aviation metalsmith, parachute rigger, radioman, aviation free gunnery instructor, navigational aids instructor, aviation electronic technician's mate, aviation ordnanceman, control tower operator, and transport airman.[40]

WAVES did not fly aircraft as pilots. The women who served as noncombat crew members and the 100 women officers trained as navigation instructors did receive 50 hours of flight time.[41] There were a few WAVES who had pilot's licenses, however, and who found ways to put their skills to use unofficially. For example, Irene N. Wirtschafter, an ensign with the Navy Supply Corps in 1944, had flying skills that did not go unused during the war. Wirtschafter (who continued with the Navy until her retirement as a

FIGURE 21.—Aviation Ordnanceman, 2nd class, Helen Wilson Looby, WAVES, stands in front of the TBF-1 Avenger. AOM 2-c Looby was one of five WAVES selected to train for this specialty. She was stationed at the Naval Air Station in Miami, Florida. (Courtesy of Helen Wilson Looby, S.I. photo 86-12197)

captain in 1976) flew some unauthorized missions for the Supply Corps. Her comment was, "We had a job to do and war to win, either I flew or supplies didn't get transferred properly; besides, I loved to fly."[42] Wirtschafter and other women like her had a certain daring. They were willing to take risks if it meant doing something important for their country and the Navy which they loved, but there are no official records of such service, no mention of hours flown in these women's log books. Ultimately that meant that their example and their courage would be officially forgotten.[43]

A WAVE differed officially from a male officer or enlisted man in that women originally were restricted to billets in the continental United States.[44] Other occupational discrepancies were also involved; for example, it was considered a critical test for all parachute riggers to make a jump with one of their own packs. However, at first, female parachute riggers did not have to do this. It was a source of considerable irritation among the men, until the women too were required to jump.

The inevitable tensions were there when the first WAVES entered service. *Time* magazine reported in June 1943 that

airmen had their fingers crossed when WAVES tower operators were proposed. They doubted they could master complex regulations, charts, procedures, meterological and weather skills (The WAACs have shied away from assigning women to the occupation), and were suspicious of how the women would bear up under control tower pressures.[45]

The first class of 20 women did an outstanding job, and the Navy began actively recruiting women for the position. The only real difficulty these women encountered (apart from a certain chauvinism) was climbing the tower ladders in their uniform skirts, a problem remedied by making pants regulation wear.

The pay and training received by the WAVES was identical to the men's. Monthly base pay for an apprentice seaman was $50; seaman, 2nd class, $54; seaman, 1st class, $66; and so forth up to chief petty officer, $138.[46] The initial course was not coeducational. WAVE officers were trained at the Naval Reserve Midshipmen's School run at Smith College in Northampton, Massachusetts. Recruit training was held at the U.S. Naval Training School (Women's Reserve), which was run at Hunter College in New York City. All additional training, however, was coeducational. The opening to women of the five aviation specialist schools on 1 February 1943 was particularly important. It was this action that resulted in integrated classrooms throughout the Navy.

The WAVES had two sister organizations. The Marine Corps Women's Reserve (MCWR), the last of the

FIGURE 22.—The five WAVES selected for the aviation ordnance specialty learn aircraft gun maintenance skills. (Courtesy of Helen Wilson Looby, S.I. photo 86-12193)

military women's corps to be formed, was not started until 13 February 1943, nearly seven months after the program had been authorized by Congress. The MCWR depended upon the WAVES organization for help starting up. Nineteen WAVE officers transferred to the Marine Corps, and the first women marines trained at WAVE schools. Initially half of the MCWR were put in aviation units, although later this number dropped to approximately one third. These women were principally aerographers, parachute riggers, control tower operators, and maintenance personnel. At the Cherry Point, North Carolina, Marine Corps Air Station, 90 percent of the parachute riggers and 80 percent of the control tower operators were women. Other billets with more limited enrollment were serial gunnery and Link trainer instructors.[47]

One of the reasons Marine women were so active in aviation was that their director, Ruth Cheney Streeter, held a commercial pilot's license. Comfortable with the world of aviation, Streeter, like Joy Bright Hancock of the Navy, was a valuable role model and helped encourage Marine officials to allow women to work in these specialties. The MCWR was deliberately kept small. Initial recruiting goals were for 500 officers and 6,000 enlisted women, although this was eventually raised to 1,000 and 18,000 respectively. The actual total of women in Marine uniforms in October 1945 was 8,500.[48]

The second "sister" was the Coast Guard's SPARs, which was organized in November 1942. Their name came from the Coast Guard motto: "Semper Paratus— Always Ready." The corps mirrored the WAVES in everything from bureaucracy to uniform. Its goal was to release men for sea duty, so although its enlisted women performed the same sorts of tasks as WAVES, including aeronautical activities, there was not a great emphasis on women serving the Coast Guard in aviation roles. Like the WAVES and the MCWR, no

SPARs were allowed to pilot aircraft. SPAR officers received billets in aviation ordnance, aerological engineering, navigation instruction, and aviation gunnery instruction. In terms of enrollment, SPAR leadership, including Director Dorothy Stratton, recognized that the Coast Guard was competing with all the other services (especially the Navy) for recruits, and thus they set very modest goals of 1,000 officers and 10,000 enlisted women—goals that were achieved.[49]

On the first anniversary of the WAVES, 27,000 Navy women were on duty; on the second, 72,350; and by the third in July 1945, 86,000 women were serving in the United States and the territories of Alaska and Hawaii. More than one quarter of the WAVES, 23,000 women, were involved in some aspect of naval aviation. One month later, in August 1945, the Navy announced its demobilization plans for women. Demobilization was mandated by law, but in fact, the Navy did not want to lose any of its female personnel. At the same time, it did need to achieve a more moderate peacetime level of operations. Although training for reserve personnel was temporarily discontinued, the women had been successfully integrated into the system, providing the historical evidence and justification for their continued presence in these "nontraditional" fields.[50]

From the beginning of the war effort, the Army Air Force leadership sought to incorporate women into its program in assignments that were not limited to the ground. General Arnold wanted to use women pilots as pilots, and thus in 1942 he became the critical factor in the efforts leading ultimately to the creation of a paramilitary organization called the Women's Airforce Service Pilots (WASP). By adopting the idea of hiring women as civilians, Arnold neatly sidestepped the limitations of the women's reserve legislation as articulated by Congress. He could not, of course, avoid confronting public sentiment, which generally was not favorably disposed toward allowing women to be military pilots. Notwithstanding this problem, he thought the use of women by the AAF would help to spur his troops to greater achievements and to dramatize the urgent need for the mobilization of other dedicated, highly motivated pilots.

Furthermore, in light of the fact that women had effectively been closed out of the CPTP in June 1941, he assumed there were female pilots who would be grateful simply to have permission to participate. By capitalizing on that gratitude, Arnold hoped he could recruit and deploy women to accomplish many of the essential but routine aviation tasks such as ferrying. This, of course, would free additional men for the higher priority wartime tasks of combat flying. Arnold could see that the potential benefits—benefits that were being realized in countries such as Great Britain and the Soviet Union—were tantalizing.[51]

This type of program would also enable Arnold to take advantage of the leadership offered by such women as Jacqueline Cochran and Nancy Love. Further, it was an ingenious way of demonstrating that the AAF really deserved its own independent women's corps. If the existing legislation could be changed to permit this, then Arnold would have a well-established group ready for action. Unlike the Marine Corps or the Coast Guard, both of which depended on the WAVES, Arnold would not have to "steal" leadership from the WAC to implement his program, nor would he experience any great time delay because of transition procedures.

Arnold had not expected the simultaneous creation of *two* groups of women pilots. The AAF leadership had been considering various plans starting in 1941, but no consensus existed as to what might be created, except that in the summer of 1942 Arnold had ordered General Olds of the Air Transport Command not to do anything with regard to women pilots until Jacqueline Cochran returned from England, where she was responsible for American women in the ATA. Olds had been the most eager to use women pilots, and Arnold probably suspected that Olds might try to go ahead and hire them. This would have put Arnold in a most difficult situation, as he had promised Cochran—under pressure from the White House—that she would direct any women's program instituted by the AAF. Arnold knew that Olds disagreed with Cochran's plan for a training program—he also had some misgivings about it himself—but he did not want to risk snubbing Cochran and incurring the disfavor of both the White House and the War Department over a matter that represented only one small facet of the entire aviation effort in the war.[52]

With the establishment of the Air Transport Command on 9 March 1942, the Army Air Corps Air Ferrying Command was reconstituted as the Ferrying Division of the ATC. The new Command consolidated a variety of programs serving the Army's transportation needs, and was firmly ensconced under General Arnold's authority. In addition to the Ferrying Division, the ATC included six foreign "wings." At the time of the reorganization, Brigadier General Harold George was in charge of the Air Ferrying Command, having replaced General Olds, who became head of the Second Air Force. With the reorganization, General George was placed in command of the entire ATC and Colonel William H. Tunner assumed responsibility for the Ferrying Division. This reorganization, which had also brought Nancy and Robert Love into the ATC, would be critical to the creation of a women pilots' program.[53]

Colonel Tunner had not been a part of the earlier discussions about women pilots, nor had he realized that their use by the military was a possibility. He was surprised to learn that the wife of his colleague Robert Love was an active pilot with a commercial rating; furthermore, Love indicated that she had more than a passing interest in the idea of using women pilots to ferry military aircraft. Truly hampered by a chronic shortage of qualified pilots, Colonel Tunner met with Nancy Love. Once again, Love presented her plan for an elite corps of women flyers, and at last she found an enthusiastic patron. Tunner conferred with General George, who immediately transferred Love to Washington to draft a proposal. A week later, on 18 June 1942, Tunner submitted a modified version of Nancy Love's plan to General George.

The proposal compromised some of Love's original ideas. Instead of choosing women on the basis of equal standards with the men of the Ferrying Division, female candidates would have to pass a more rigorous set of requirements.

	Men	Women
Age	19-45	21-35
Education	3 years high school	high school graduate
Required flight time	200 hours	500 hours*

*At least 50 hours flown in previous year.

Other more substantial changes were yet to come. Love's proposal assumed the women would follow the same path already established for the men of the Ferrying Division: they would be hired as civilians, and following a 90-day trial period they would be commissioned into the AAF. Just before Tunner submitted Love's plan, the WAAC legislation was passed in Congress. Love and Tunner met with WAAC director Oveta Hobby to discuss the plan for women pilots. Hobby thought the program was an outstanding idea, but unfortunately, the three did not immediately realize that the WAAC legislation did not allow for flying officers and flight pay in the women's corps: women pilots could not be legally commissioned in the AAF.[54]

When this problem became apparent, Love and Tunner recognized the great difficulties that would be encountered in any attempt to persuade Congress to amend the WAAC legislation. They were aware that the WAVE proposal, then before Congress, was experiencing substantial difficulties because it included provision for women pilots. Gambling that legislation would eventually be passed to permit women pilots in the military, especially once women demonstrated their abilities, Love proposed they go ahead and hire the women as civilians. She stipulated, however, that the

ATC had to make clear to the women that their status was temporary. She also asked that the ATC exert every effort to get these women commissioned.

Love recognized that her pilots would be scrutinized because of their uncertain future, so she further stiffened the requirements for women candidates. Now a female pilot had to have a CAA 200-horsepower rating and two letters of recommendation to be considered. Love wanted "blameless" types with exemplary personal conduct. Finally, she proposed that the women's salaries be set at $250 per month, $130 less than the men civilian pilots received. The reason for this was that the women would be flying only small trainer and liaison-type aircraft.[55]

The final written version of the plan for the Women's Auxiliary Ferry Squadron was sent to Arnold on 3 September 1942, along with current CAA statistics on women pilots, which Arnold had previously asked Love to update. Two days later, Eleanor Roosevelt restated the argument for women pilots in her "My Day" column. On 5 September 1942, General George decided to proceed with the program. According to Jacqueline Cochran, Arnold knew nothing of the WAFS plan after seeing it in July because there did not appear to be any official record of authorization. It seems preposterous to assume that Arnold did not "know" anything about George's decision to implement the WAFS program. The ATC had formally named Nancy Love to be director and it had sent telegrams to prospective female candidates to join the program. Finally, the date for the public announcement of the creation of the WAFS was set for 10 September; the announcement was to be made by Arnold from his office.[56]

Two things of interest occurred at this time. The first was the delayed departure of Jacqueline Cochran from England. As she was preparing to depart for the United States, General S.H. Frank of the Eighth Air Force had Cochran called from the airport, ostensibly on an urgent matter. The matter turned out to be a thank-you dinner, but it held up Cochran for three days. She suggested that this might have been a deliberate ploy by the ATC or Nancy Love to keep her from interfering at the last minute in the events unfolding in Washington.[57] It probably was not, but it did effectively prevent her playing any role in the week's scenario.

The second odd turn of events was the absence of General Arnold on 10 September. When Love and George arrived at Arnold's office for the morning press conference, they were told that he had been called out of town; the Secretary of War, Henry L. Stimson, would make the announcement. It is possible that Arnold foresaw he would have great difficulty explaining to Cochran the following morning (he and Cochran had an appointment on 11 September 1942) why he had reneged on his commitment to make her the director of any women's flying program that the AAF might establish. If Stimson, on behalf of Roosevelt, made the announcement, it would deflect Cochran's irritation from Arnold onto George and Love. It would also prevent her going over Arnold's head to complain to the Roosevelts.[58]

Arnold's equivocation resulted in the creation of two women's flying groups. He conceded that he had not fulfilled his earlier promise to Cochran, and therefore he agreed to start the women's training program under her leadership. This training program was known as the Women's Flying Training Detachment (WFTD, sometimes called "Woofteds"). Eventually it was fused with Love's Women's Auxiliary Ferry Squadron (WAFS) to become the Women's Air Force Service Pilots (WASP), with Cochran as overall director and Love as her subordinate running the WAFS.

The WAFS and the WFTD, as well as their successor organization, the WASP, comprise a vital component in the story of women in military aviation. While these women technically remained civilians and were drastically outnumbered by women in the regular military groups, their experiences present a paradigm for the service of World War II United States military women as a whole. Examples drawn from this elite organization vividly and accurately illustrate the scope of experience of their military sisters.

4. Nieces of Uncle Sam:
The Women's Airforce Service Pilots

On through the storm and the sun
Fly on till our mission is done
From factory to base,
Let the WASPs set the pace.

from "WASP Song," Loes Monk, 43-W-8

Beginning in September 1942, there was a 10-month period of cautious and superficial calm between the leaders of the Women's Auxiliary Ferry Squadron (WAFS) and the Women's Flying Training Detachment (WFTD). Nancy Love, director of the WAFS, and Jacqueline Cochran, director of the WFTD, represented two very different personal styles. Each drew partisan support from interested individuals who were in sympathy with the vision or character of one or the other of the two leaders. Love evoked loyalty and respect from her small group of women pilots. A leader by example, she was an active participant in her cadre of talented flyers. Cochran was, by contrast, a born administrator, skilled in achieving her ends by personal influence and negotiation.

There were those within the Pentagon and on Capitol Hill who were inspired by Cochran's political skills, and others who saw Love as the most gifted leader of women in the field. Out of the conflicting attitudes emerged a power struggle. This battle was initiated by Cochran, who, some thought, aspired to a position of complete control over all women's military flying programs.

The first evidence of this is an angry memorandum sent to Colonel William Tunner from Captain James Teague of the Air Corps on 22 September 1942. Teague wrote:

... she [Cochran] made it quite clear that she considered herself the only person who could efficiently be in charge of the Women Ferry Pilots. This was all done by innuendo, and at no time did she actually express this thought in words of one syllable.[1]

Teague voiced further complaints about Cochran because she apparently wanted her own opinions to supercede Nancy Love's authority in determining the final selection of WAFS members. He concluded his memorandum:

We are in this position: Miss Cochran, as far as the public is concerned, is coming to us and bringing us women who have been

trained, and we should be appreciative. I, on the other hand, fear the Greeks bearing gifts. I have discussed this with Capt. Tucker and he agrees with me that some method should be found by which those people who are in authority above us should be told exactly what our attitude is, and a clear line of demarkation [sic] drawn now. I am afraid that if we let this thing go too long, Miss Cochran will take inch by inch and try to move in on us. I don't believe I am exaggerating the extent to which she will go.[2]

Captain Teague's memorandum was only the earliest piece of written evidence for the conflict that surrounded the emerging Love-Cochran rivalry. Other documentation confirms that various military leaders took sides on this issue, generally supporting one woman or the other. It is likely, however, that these partisan opinions were influenced as much by personal attitudes regarding women's participation in the military as by any assessment of the programs or leadership style associated with either of the two women. There was little high-level support in the military for a female pilots program, but also little public opposition. Many of those with strong views believed the issue was so controversial that to vocalize their objections would have a negative impact on their job.[3]

For a variety of reasons, Tunner, George, Olds, and Arnold all wanted women in their programs. Countering this enthusiasm, however, was the constant scrutiny of every activity that women pilots engaged in. For example, in July 1943 the two paramilitary programs for women pilots came under Senate investigation by the Truman Committee. Confidential memoranda between Julius Amberg (Special Assistant to the Secretary of War), Lt. Colonel Miles H. Knowles, Colonel G.A. Brownell, Maj. General Barton Yount, and General Arnold during this month went back and forth raising charges, complaints, and allegations about "Cochran's program." Some of the military support for "Love's program" may have been derived from the fact that it was perceived as the lesser of two evils.

Whatever the internal problems, the ATC and the AAF did not publicly discuss the issue; however, stories such as the July 1943 *Newsweek* article entitled "Coup for Cochran," occasionally did become public, in spite of the ban on press coverage of the WAFS imposed by the War Department in October 1942. All parties directly involved were eager to treat the women as if they were in the military with the expectation that this would soon be a reality. The women pilots for their part were just as eager to participate in flying for the war effort. Applications came flooding in, making the WAFS and the WFTD the only women's groups affiliated with the military to have a surplus of recruits.

The "Original 27" WAFS were handpicked, but the invitation to apply did not guarantee a position. If the woman accepted, she would report to an AAF base (at her own expense) for a physical. If the results were satisfactory, the woman would then be asked (again, at her own expense) to report to the New Castle Air Base in Wilmington, Delaware, to undergo flight checks, examinations, and a personal interview with Nancy Love. Finally, the candidate would appear before a review board of three ATC officers who would examine

FIGURE 23.—Nancy Harkness Love, WAFS director, greets four of the "Original 27" recruits for the Women's Auxiliary Ferry Squadron as they arrive at the training base in New Castle, Delaware: (left to right) Love, Cornelia Fort, Helen Mary Clark, Teresa James, and Betty Gillies. (Courtesy of Delphine Bohn, S.I. photo 86-5616)

her dossier (Nancy Love was permitted to sit in as an ex officio member). In order to become a WAFS member, a woman had to receive the board's approval, just like any male Air Force candidate.[4]

The original WAFS were an elite corps. Among the most experienced women pilots in the nation, they were articulate, bright, and enthusiastic. They exuded an aura of athleticism, good humor, and self-confidence. They are remembered as talented individualists from a variety of different backgrounds, not as "society dames" as Cochran once called them. It is important to remember that some of the WAFS were as talented at flying as Cochran herself. It would take more than flight experience for Cochran to earn respect and the right to leadership among these women.

The ATC needed convincing that women could really fly, so they subjected them to all manner of tests and ground school classes, even when the women had more experience then their military instructors. They also introduced the women to military procedures. Finally on 20 October 1942, 40 days after the invitations to apply had gone out, WAFS began to ferry aircraft.

Piper L-4 Grasshoppers and Fairchild PT-19s were the first craft assigned. Stoically the WAFs performed the physically, if not intellectually, arduous task of transporting these trainers from the factory to the various air fields around the United States. They did extremely well during the first months, with a delivery record of 100 percent. Their performance was noticed by the ATC commanding officers, and in December 1942 General George commented: "If they can fly four-engined bombers safely after proper periods of training and preliminary work, I see no reason now why they may not get the chance."[5]

Cornelia Fort, one of the original WAFS, best expressed her colleagues' desire for acceptance:

Because there were and are so many disbelievers in women pilots, especially in their place in the army, officials wanted the best possible qualifications to go with the first experimental group. All of us realized what a spot we were on. We had to deliver the goods or else. Or else there wouldn't ever be another chance for women pilots in any part of the service.[6]

A new era had begun. For the first time, American women were flying military aircraft. To the surprise of many, they did a good job, and they provided significant service. The WAFS were proud of this, but most of their satisfaction came from the knowledge that they were able to combine an activity they loved with a challenge and a service. They also enjoyed working for Nancy Love, who was considered a model of the ideals many WAFS held. She was regarded as a great leader because she was a pilot first and an administrator second. Her willingness to fly tough missions, surprise the group by arranging for the provision of their own

grey-green uniforms, or tirelessly act as their advocate were among the many reasons accounting for her popularity. Her personal conduct was always gracious and civil. Most important, she conveyed her pride in the organization without appearing to seek personal recognition.[7]

The WAFS expanded their base of operations after the first trial months, and Love named several WAFS as commanding officers (COs) at the new bases of operations. When Love was transferred to ATC headquarters in Cincinnati, Betty Gillies became CO at New Castle. Delphine Bohn headed a WAFS flight squadron at Love Field, Dallas, Texas; Barbara Donahue (Ross) at Romulus, Michigan; and Barbara Erickson (London) at Long Beach, California. Farmingdale, Long Island (at the Republic Plant), and some other locations were added later. All of these COs performed well, but Barbara Erickson, later WASP Commander at Long Beach, California, deserves special notice. In March 1944, Erickson, who had flown tirelessly, was awarded the Air Force Medal by General Arnold and given a citation signed by President Roosevelt. The marathon mission for which she earned this honor involved making four 2000-mile flights in three different aircraft within a five-day period.

The WAFS began to grow larger once Jacqueline Cochran's training program was under way. On 24 April 1943, the first WFTD class graduated and all 23

FIGURE 24.—Gertrude Meserve (left) and Teresa James (right), WAFS members, wish each other well before the start of a ferrying mission. Both were part of the WAFS squadron based at the Republic Aircraft Plant in Farmingdale, Long Island. (Courtesy Gertrude S. Tubbs, S.I. photo 85-17017)

FIGURE 25.—Helen Richey (left), America's first commercial airline pilot, and Dorothy Colburn (right) are returning to their home base at the Republic Aircraft plant in Farmingdale, Long Island, after a ferrying flight. (Courtesy of Gertrude S. Tubbs, S.I. photo 85-16994)

members reported to the WAFS. They were followed by 43 graduates in May and 38 in early July. The fourth graduating class with its 112 women pilots reported to ATC bases in August 1943, and the WAFS were on the way to operating as a full-strength Ferrying Squadron. Approximately 150 more graduates would be assigned to the WAFS before the end of 1943.

By this time Cochran had convinced the AAF to assign women pilots to various other flying duties throughout the training commands and numbered air forces. In August 1943 the AAF reorganized the command structure of the women pilots. The WAFS, who before August were a separate entity under the Ferrying Division of the ATC, were placed under the control of the Director of Women Pilots. Cochran was appointed to this new post, and although Love remained in charge of the women pilots from the Ferrying Division, she now reported directly to Cochran in the Pentagon on matters relating to assignment and disposition of personnel. Cochran was still in charge of training operations although the WFTD name was dropped. The new organization was named the Women's Airforce Service Pilots or WASP. This was still a civilian group but the AAF had created a single hierarchical unit through which it could experiment with the deployment of women pilots.

48

FIGURE 26 (left).—Betty Gillies (left) and Nancy Harkness Love (right) of the WAFS, stand in front of the B-17G before their transatlantic ferry trip attempt to England in September 1943. Their mission was grounded by General Arnold before they left Gander, Newfoundland, on the final leg of the trip. (Courtesy of Gertrude S. Tubbs, S.I. photo 85-11125)

FIGURE 27 (below).—Women's Airforce Service Pilots (WASP) Betsy Ferguson (left) and Florence Miller (right) prepare for a flight in a North American AT-6 trainer aircraft. (Courtesy of Delphine Bohn, S.I. photo 86-5599)

By June 1944—D-Day in Europe—the AAF began to retrench, closing down most of the pilot-training bases. This brought about a major ferrying job, which required the services of hundreds of pilots to move surplus training aircraft from these bases to storage and disposition areas throughout the country. The decision was made to transfer to the training command all WASPs assigned to the WAFS section except for those pilots who were qualified to fly pursuit aircraft. (The WAFS were thus subsumed under the WASPs; only the "Original 27" maintained their WAFS identification—despite the reorganization.) There were 123 women who qualified as Class 3 pilots, capable of flying combat aircraft such as the P-47 Thunderbolt fighter to ports of embarkation in support of the invasion of Europe. All of these women possessed instrument ratings, i.e., they were able to fly on instruments alone; 98 percent had twin-engine ratings and 80 percent had both single- and twin-engine fighter aircraft ratings. Five of them, Nancy Love, Betty Gillies, Barbara Erickson, Dora Dougherty, and Dorothea Johnson, were qualified at the highest level, Class 5. Dougherty and Johnson later qualified to fly the B-29 bomber aircraft.

Many of the original WAFS were not interested in being a part of "Cochran's program," and despite the name and program change, they never considered themselves WASPs. Even today, some of these women identify themselves as WAFS, express preference for their original uniform and loyalty to Love rather than to Cochran. The young women who learned to fly the "Army Way" through Cochran's training program were less affected by such tensions. Jacqueline Cochran was their hero and many hoped to emulate her remarkable personal achievements.

In the spring of 1943, Cochran told the first class of trainees that this was "the greatest opportunity ever offered women pilots anywhere in the world."[8] The Women's Flying Training Detachment (WFTD) was first based at the Houston, Texas, Municipal Airport. (The later designation for the Houston phase of the program was the 319th Army Air Forces Flying Training Detachment—AAFFTD.) There was a deluge of applications for this program (in the end, Cochran received more than 25,000),[9] but the requirements were stiff. Like the WAFS, applicants had to be American citizens between 21 and 35 years old. The women had to be at least five feet tall (later five feet, four inches) and pass the flight surgeon's physical. The difference between the WFTD and the WAFS was in the extent and level of previous flight time required for admission. The WAFS demanded 500 hours, a 200-horsepower rating, and a commercial license. The training program initially required 200 hours without regard to horsepower.

Following the outbreak of the war, obtaining a license became quite difficult because of the ban on flying within 100 miles of the United States border, the extra security clearance requirements on cross-country flying, and the shortage of civilian aircraft and instructors. For these reasons, the requirement of a valid license was dropped before the first class was filled. The required hours were subsequently reduced to 75 by January 1943, and ended up with a minimum requirement of only 35 hours.[10]

The other important aspect of the selection process was the personal interview with Cochran or one of her representatives. The interview was used to assess the candidate's personality, her stability, and various aspects of her background that might be indicative of her future performance under stress. Given the physiological requirements and the limited number of women who were pilots, the women who were finally selected tended to be fairly similar.[11] They were certainly kindred souls when it came to flying. This was manifest in the feeling that the recognition of their sister pilots was more important than the approval of their male flight instructors. The entire WASP experience gave them a unique self-confidence, resulting in the ability to persevere through all kinds of difficult and unusual situations during the war and, naturally, later on in their lives.

The personal interview also allowed Cochran to determine the race of the applicants. The program had only limited minority participation: two Chinese-Americans, one Native American, and a few Jewish women. There were no black women, although several applied.[12] Cochran was remarkably candid in her autobiography about her treatment of these applicants. She hoped that they would fail the preliminary examination, so she could always claim fair treatment.

The first black woman (unidentified by Cochran) to successfully reach the final stages of selection had a rather unusual reception. Cochran made a special point of interviewing her. At that meeting Cochran says that she made a strong case for the problems of integration in military life; further, she stressed the program's experimental status. According to Cochran the young woman understood the situation and withdrew her application. The barriers a black woman (and all minority women) faced were three-fold—race, class, and sex—and each militated against her participation. Given the climate of the day, Cochran probably was correct in assessing how difficult it would have been to get both the military bureaucracy and Congress to overcome their bias against blacks as well as against women. She believed that to interject the race question into the project might well have spelled its demise.[13]

The discrimination and the profound difficulties,

even dangers, that would have had to be faced by a lone black woman pilot ferrying aircraft to isolated bases all over the country were very real. Cochran sensed the true sentiment of the military towards the participation of blacks. Just by acknowledging the issue of racism, Cochran was more conscious of it than the majority of white Americans. However, her actions continued to reflect the majority sentiment.[14]

When the WFTD candidates who had been selected arrived at Houston (at their own expense) in October 1942, they entered a program that was still being created. The women boarded in hotels and started classes run by a contractor, Aviation Enterprises. The goal was to teach the women to fly Army trainer aircraft. The first approved curriculum had 115 hours of flight time, 20 hours in a Link Trainer, and 180 hours of ground school. The trainees first flew in liaison aircraft, such as Piper Cubs and Taylorcrafts. In the second phase they flew in military training planes, such as the PT-19 and the BT-13. The last 15 hours were spent in advanced trainer aircraft, such as the AT-6 and the AT-17. The trainees wore large coveralls designed for men (affectionately called Zoot suits) and fleece-lined leather flying jackets. No provisions for meals and formal uniforms were made, which meant that often the women did without. In other ways the group was expected to be just like the military, and their curriculum included physical training, calisthenics, and marching.[15]

The women of this first crew had initially expected their training to last only a few weeks. It turned out to be a demanding five-month course, but the spirit of these women was indomitable. On the long bus rides between Houston and the airport, they sang and they laughed. Bound together by a spirit of comraderie, this class is best characterized by their altered rendition of George M. Cohan's "Yankee Doodle Dandy":

We are Yankee Doodle Pilots
Yankee Doodle, do or die!
Real live nieces of our Uncle Sam,
Born with a yearning to fly.
Keep in step to all our classes
March to flight line with our pals
Yankee Doodle came to Texas
Just to fly the PT's!
We are those Yankee Doodle Gals!

Gradually, conditions improved and the program became more organized. The second class arrived in December 1942. They were still located in Houston, but at least the contractor had opened a mess hall, to which the women were expected to march, Army style. During their tenure, housing was consolidated to a few large blocks of cottages and motel rooms. The curriculum underwent revision and extension, until it was much like the AAF program for male cadets and lasted 22 1/2 weeks. The chart below indicates the training schedule.

PRIMARY (50 hours)	
Fundamentals of flying	46
Navigation	4
BASIC (70 hours)	
Transition (to BTs)	30
Instruments	20
Navigation (Day 18, Night 2)	20
ADVANCED PHASE (60 hours)	
Transition to AT-6	10
Transition to twin engine (AT-17 or AT-10)	20
Navigation (Day 18, Night 2)	20
Instruments	10

The first three intakes and half of the new fourth class were based at Howard Hughes Field, Houston Municipal Airport. There were just 22 various civilian and military training aircraft available to them there, but that did not at first affect the capability of the school. More serious was the winter weather along the Gulf coast, especially the fog, which resulted in a perpetual scramble for the women to log the specified number of hours needed to complete the course. This required constant rescheduling and flying on Sunday, the only free day. Despite the difficulties, the women were doing exceedingly well, and the decision was made to increase dramatically the number of students. The original plan for 396 students was increased first to 700 and then to 1,000 in 1944.[16]

This decision made it imperative to seek new facilities. From the start, Cochran worked to align the WFTD (and later the WASP) completely with the Army training program. Although following procedure and method was relatively easy, using military facilities was not. Cochran was successful in getting an entire military base for the training of women pilots. The new base, Avenger Field in Sweetwater, Texas, was being used to train Canadian cadets, and Mrs. Leni Leah Deaton, Cochran's chief administrative officer, found them still on the field when she arrived to prepare for the WFTD move. Deaton, who had set up the operation in Houston, managed the enormous task of moving half of the existing unit to Sweetwater, while also preparing for the integration of a new class every month through 1944. Deaton had been a Red Cross swimming administrator for many years and brought with her a broad knowledge of how to manage large programs. By April 1943 the WFTD had the field to themselves. With the move to Sweetwater, the group became the 319th Army Air Force Flying Training Detachment. They finally had adequate room for the training program run the Army way.

Not everyone cared about the Army way. A few went so far as to say that the military aspects of the program were something to be endured. Toby Felker of the 44-W-2 (i.e., the second class of women pilots in 1944) proclaimed that, to her, what really mattered was the opportunity to fly. "All sorts of very interesting things happened because these women who liked to fly—who wanted to fly—were going to do so, using any means they could. I really felt that was an exciting part of it. None[17] of us really cared very much for the military regiment [sic] . . . but that was just one of the things you had to do."[18]

Flying was, of course, the main activity during training, "their single-minded consuming passion," as Doris Brinker Tanner described it. Life at Sweetwater marked the change to Army-style living with barracks, footlockers, and shared bathroom facilities. The flying curriculum was gradually expanded, partly in response to the applicants' diminishing flight experience (200 hours had been lowered to 100, to 75, and finally to 35 hours) and age (from 21 to 18½). The final program was a 30-week course of flight training and ground school classes. The flight training consisted of 210 hours of aerial instruction. Ground school involved 560 hours divided among various programs in military training, academic studies, aircraft equipment maintenance, physical education, and medical training. Academic work included mathematics, physics, navigation, the principles of flight, engines and propellers, code, instrument training, and communication. Examinations were inevitable and tensions were high for both ground and aerial tests.

Contrary to expectations, the elimination rate for the women proved to be about the same as for AAF male cadets. Of the 1,830 women applicants accepted (out of 25,000 inquiries), 30.7 percent were eliminated due to flying deficiencies and 2.2 percent for other reasons. Eight percent passed but resigned before assignment. This resulted in the final tally of 1,074 graduates, or 58.7 percent of the total who had been accepted.[19] Most of the "wash outs" occurred in the primary phase (although it was not unusual for candidates to experience difficulty when making the transition from light aircraft to the PT-17 or PT-19). When comparing the success of male and female student pilots, it is important to note that the AAF sometimes deliberately failed candidates (by altering requirements) in a particular class, if there were fewer pilot combat casualties than anticipated.[20]

Some women did not complete the training program because of accidents. Of the 38 fatalities in the overall WAFS/WFTD/WASP program, 11 occurred during training.[21] Aside from the obvious aspects of the loss, the hidden tragedy of these accidents was that there was no insurance, no death benefits, in short, no official recognition of service. As civilians, the women began their training aware of the risks as well as the benefits, but gradually, through the training program, they became thoroughly indoctrinated with the concept of service for their country, just as the male cadets were. When a woman died, the other women felt betrayed that the government was incapable of expressing condolence through some formal gesture of recognition to the survivors. Their expectation that this gesture should be made, in terms of some sort of financial assistance or insurance, was in accord with the values of President Roosevelt, the New Deal, and, increasingly, American society at large. In being treated and trained as equals with men, they had come to expect, at the least, equal recognition in the form of comparable death benefits.

Graduation brought a number of options for the women who succeeded. At first, there was only one task—that of ferrying trainer and liaison aircraft. Following the August 1943 power shake-up between Love and Cochran, each successive graduating class gradually expanded their range of tasks and responsibilities. The next major duty to be undertaken was target towing at Camp Davis, North Carolina. This meant flying with a long strip of fabric attached to the airplane by a long tow line, in order to provide anti-aircraft gunnery practice. At first the women were not welcome. Twenty-five were assigned to test fly Piper Cubs, doing work incongruous with their training, until Cochran visited the base. Finally, the women were put on the target towing project.

The condition of the aircraft at Camp Davis was considered to be quite poor. Priority for parts and supplies, not to mention new airplanes, was held by AAF units in combat. One woman in the Camp Davis program was killed in an accident caused by contaminated fuel, an incident that was strongly rumored to have been the result of sabotage. Cochran denied this, and investigations resulted in a plausible alternative explanation of equipment failure. Years later, however, Cochran admitted that sugar had been found in fuel tanks and that other women had died because of sabotaged aircraft. This information had been suppressed for fear that adverse publicity would end the entire women's program.[22]

Gradually, the base commander, admittedly prejudiced against women, came to accept the group and acknowledge their skills. They were used for searchlight tracking practice and mock strafing missions. Soon thereafter, classes were sent to many bases to fly radar missions and remote-control drone airplanes.[23]

A few WASPs were sent to advanced schools after graduation. Seventeen of them from classes 43-W-5

FIGURE 28.—Evelyn Sharp, WASP, stands in front of the Douglas C-47 airplane of which she was the ferry pilot. Sharp was one of the 38 WASPs killed in service to the United States. She died when her Lockheed P-38 airplane crash-landed during a ferry mission. (Courtesy Delphine Bohn, S.I. photo 86-5600)

and 43-W-6 went to Lockbourne Army Air Field in Ohio to learn to fly B-17s. When they had successfully completed the course, 13 of them reported to operational assignments, flying gunnery and tow targets for aerial and ground gunnery practice. A second group from these two classes transferred to Dodge City, Kansas, to train on the B-26 bomber, known to be a difficult craft to handle. In the case of the B-26 and the B-29, WASPs were deliberately used to counter the undeserved discredit then given these aircraft by the men. The theory that "if a woman flies it, it must be safe" proved effective, and it only took a few flights

(along with an extended wingspan) to have the image of these airplanes transformed into "safe" and "good" aircraft according to the male pilot "grapevine."[24]

WASPs also worked as administrative pilots, assigned to different groups such as the weather wing, which worked with the Air Weather Service. They flew cargo runs and often flight tested newly delivered aircraft. One WASP, Ann Carl, became the first woman (and one of the first pilots in general) to test fly the Bell YP-59, America's first jet fighter.[25] When the WASP program ended in December 1944, the 916 WASPs on active duty were assigned primarily to the training

FIGURE 29.—Sixteen of the 17 WASPs qualified to fly the B-17 bomber. (Courtesy Lt. Colonel Yvonne C. Pateman, USAF (Ret.))

command and the air transport command. Eighty women worked with the Second Air Force, with the remainder scattered among various bases and commands.[26]

Despite the variety of their assignments, most WASPs exhibited little awareness of other women in the military. Few recall having much contact with other women apart from the nurses. Periodically, however, hints of a bit of inter-service rivalry would surface. For example, the verse of a WASP favorite song asks facetiously if the listener would rather be a WAC:

A WAC may be an officer
With bright bars that shine,
Her olive green and everything looks fine.

She's very proud of the name she bears,
As for you, you don't want her cares;
Her olive green was never meant for you,
You want the Santiago Blue.

Santiago Blue referred to the attractive new uniforms that arrived in April 1944. In 1943, the AAF Quartermaster General had proposed to Cochran that the women pilots wear uniforms based on surplus WAC and Army Nurse uniform supplies. She summarily rejected this idea and arranged for a New York designer to create a new outfit, which was finally approved by General of the Army, George C. Marshall.

The women were glad finally to have something to replace the awkward, ill-fitting and very unofficial-

FIGURE 30.—Ann Carl, WASP, was the first American woman to pilot a jet aircraft. (S.I. photo A45978)

looking outfits of men's trousers and shirts because, at least in part, it was another step towards professional recognition. The uniforms were an important symbol of the WASP program and they helped the wearers to feel more a part of the military and of the AAF. The response was also favorable from base commanders, military police, and other personnel who found it desirable to have the women suited up in clearly recognizable uniforms. Cochran had demonstrated her authority once more: by keeping her group visibly distinguishable from the WACs, she had made it more difficult for the WASPs to be subsumed under Oveta Hobby's control.[27]

New symbols do not always lead to the establish-

ment of the particular new reality that is desired. For the WASPs, transformation into formal military status was the critical step to be taken, but Congress was not as easily persuaded of this change as General Marshall had been in the matter of a new uniform. By the summer of 1944, Arnold was absolutely convinced that the WASP program should be continued. It was accomplishing everything he had originally hoped for, and thus he began to submit proposals to Congress for legislation authorizing the incorporation of the WASP program into the AAF. Congress, on the other hand, did not envision an expanding role for women pilots in the AAF, but rather a post-war return to normalcy—an AAF without women pilots. The emotional side of the issue was further influenced by a series of editorials and letters protesting the AAF failure to utilize the abilities of recently discharged CAA flight instructors.

A congressional investigation, chaired by Robert Ramspeck, was formed. Preliminary report comments were not positive:

FIGURE 31.—Jacqueline Cochran (middle row, far left), General H.H. Arnold (next to Cochran), and Nancy Harkness Love (behind Cochran) review WASP graduation proceedings in March 1944. This is a rare photograph, showing all three of the principal figures responsible for women in military aviation together. (Courtesy of Gertrude S. Tubbs and the Fort Worth *Star-Telegram*, S.I. photo 85-11122)

To recruit teenaged school girls, stenographers, clerks, beauticians, housewives and factory workers to pilot the military planes of this government is as startling as it is invalid. The militarization of Cochran's WASPs is not necessary or desirable; the present program should be immediately and sharply curtailed.[28]

The committee was acting on House Resolution 4219, proposed by Congressman John Costello of Los Angeles. The resolution was designed to commission the WASP in the AAF with full benefits. In response to questions put during the inquiry, Nancy Love and General Tunner compiled extensive statistics and reports from commanding officers and other personnel who had worked with the WASP. This study showed that the stereotypes about women pilots were incorrect and unfair. WASPs were not accorded preferential treatment in assignments, and if resentment existed among male pilots, it was confined to a limited (but vocal) minority whose primary interest appeared not to be the most efficient operation of the ferrying division, but rather the preservation of an all-male environment.[29]

These facts were supported by the testimony of numerous individuals, most importantly, General Arnold. In executive session Arnold explained that, in the matter of training, he preferred the WASPs over the regular civilian instructors for reasons of ability and attitude. Ultimately, however, the AAF could, according to Arnold, absorb both the WASPs and the CAA flight instructors, although the men might be used in different capacities.[30]

Cochran complicated the situation by refusing to compromise certain aspects of her program. Even though there was a favorable response in Congress toward keeping the WASPs who were already trained, there was no support for continuing the training program. It was on this point that Cochran balked. She insisted on a continuing program and on an organization separate from the WACs, because of her unwillingness to become subordinate to Oveta Hobby. Most observers at the time suggest that had Cochran been willing to attach the WASPs to the WAC, the program likely would have been militarized. Her refusal to compromise on this issue generated a high level of publicity and assured that the status of women in the military would remain controversial. Some men who had been civilian instructors were quick to highlight every criticism, including the erroneous assumption that women pilots were grounded each month for their menstrual period, thus having fewer hours of flight time than a man. The press openly criticized the women, calling the WASPs "glamour girls" who weren't legitimate pilots.[31]

Arnold's position supported Cochran: he had previously stated that the women pilots should not be incorporated into the WAC because of the "need for undivided and administrative and functional control which would not be possible if the WAAF [sic] was serving two masters, i.e., the WAAC and the AAF."[32] It should be remembered that Arnold was involved in breaking the ground to separate the AAF from the Army, and further, that he had originally proposed a separate corps of Air-WACs. For Arnold, what counted was the fact that the women were performing their jobs well. But Arnold's attempts at using the women as a means of motivating his male pilots had backfired. The men were angry over attempts to make them feel guilty and to shame them into performing aspects of their jobs they thought were risky.[33] This irritation was compounded when the women earned praise and publicity for their performance of a job that went unnoticed when performed by a man. To a certain extent the men were justified in their reactions. For instance, men who had flown tow targets did not attract national media attention; this job became "newsworthy" only when some women replaced men in its performance. Sometimes in their eagerness, the women ignored the validity of protests made by men about equipment or procedures.

The debate raged on with conservative congressmen lashing out at the military for conducting social experiments. The efforts of 1,074 women flying more than 60 million miles in almost every airplane in the AAF inventory did not, in the end, matter to Congress. The bill did not pass, and orders were given to demobilize the WASPs on 20 December 1944. During the fall, rumors percolated through the ranks about ex-WASPs being rehired. On 5 October 1944 Arnold adamantly stated: "There will be no—repeat—no women pilots in any capacity in the Air Force after December 20 except Jacqueline Cochran . . . ."[34]

Many of the women were bitter. Barbara Poole wrote a scorching piece for *Flying* magazine's December 1944 edition entitled, "Requiem for the WASP":

We have spent large sums to train the WASP.[35] Now we are throwing this money away at the demand of a few thousand male pilots who were employed, until recently, in a civilian capacity on government flight programs. The curtailment of the program has thrown these pilots out of work. And now they are to get the WASPs' jobs.

The demands of the unemployed male pilots have been a little far-fetched.

"Throw the women out," they cried—meaning "and make room for us." They say, "The women can't be drafted, but we can and will be."

But that's the very reason the women should have stayed where they were. What our Army needs most, our generals tell us, is men to fight on the ground. This is a sorry state of affairs for our pilots but after all we're running a war, not an employment bureau for disgruntled flyers."[36]

While the WASPs were prevented from joining the military, the WACs, the WAVES, and the MCWR all managed to convince the legislators of their worth. The women who packed parachutes or served as flight nurses would continue to operate in the military beyond World War II, keeping alive the consciousness of women in military aviation.[37]

Before the war the aviation organizations and events in which women were involved were small, characterized by an economically, and often intellectually, elite membership. World War II changed that by institutionalizing female participation in two major centers of power in American society—industry and the military. Women pilots were a special case. It took their desire to participate and the advocacy of several key individuals, Eleanor Roosevelt in particular, to bring men to acknowledge the inherent usefulness of a talented woman pilot (especially one willing to do virtually any kind of flying). Previously, the planning of military leaders had not automatically included women pilots, but by war's end this had changed. General Arnold's letter to each WASP announcing demobilization noted that the nation *knew* it could count on women pilots for any future effort. "It is on the record," he said, "that women can fly as well as men. We will not again look upon a woman's flying organization as experimental."[38]

The women in military aviation were proud of their accomplishments. They wanted recognition and publicly acknowledged status in conjunction with their work, which is why they were so concerned about military status. Most WASPs felt that being in the military was peripheral to the *act* of flying military aircraft; to them, demobilization was devastating. Even though these women could continue to fly civilian aircraft, there was a keen sense of loss at not being able to fly military airplanes. It took 30 years for the controversy to cool sufficiently for women to petition Congress seriously again on this subject. Nevertheless, when the war ended, women in civilian and military aviation joined in the general relief and happiness that it was finally over. Like other Americans they looked forward to the "return to normal" that peacetime was supposed to bring.

Their participation at every level of society, whether university or factory, Civil Air Patrol or Ninety-Nines, WAC, WAVE, or WASP, had established the foundation for a new age in aviation for women. The exceptional leadership offered by such women as Cochran, Love, Hobby, and Hancock marked the first appearance of women in such responsible roles. They were successful, competent women who became integrated into the existing power structure by adopting the norms and expectations of the male leadership. Their acceptance into the top echelons was generally contingent upon their not challenging the existing power structure.[39]

Establishing a permanent presence for women in this arena was a difficult task because, in the eyes of American society, all of the programs for women in military aviation were experimental. Leaders who sought to overcome existing prejudices insisted on exceptional job performance from the women in their programs. The quality of a woman's work had to equal or exceed that of a man's if unreasonable biases were to be dispelled.

Although discrimination in the armed forces based on gender had lessened somewhat by the end of the war, racial segregation still prevailed. There had been little real progress in the integration of women of color into military aviation. Here, as in the rest of society, integration was viewed as too controversial for serious consideration; attempting to put it into practice was still believed to jeopardize the already tenuous status of any program for women. A change would not come until 1948 with the Integration Act.

During the war American women had demonstrated to themselves and to all United States citizens that women were capable of working in aviation occupations. A significant period of time would have to elapse before this knowledge would have its full effect. Building on their wartime experience, women expanded their aviation opportunities enormously over the next 40 years, and by 1985 it would be clear that United States women constituted a community within the aviation world instead of being isolated individuals. This is the legacy of World War II.

Postwar Era

5. Demobilization and the Postwar Transition: 1945-1949

They were girls who could not sit beside
The hearth and let go by
All the joy and pride and thrills that ride
With rovers of the sky

from "To the Ninety-Nine," Louis De Jean

Women's participation in aviation, institutionalized during World War II, was somewhat tenuous during the immediate postwar period. Wartime expectations of what the return to peace would mean for America were quite different from what actually happened. The war seemed to have fostered an optimistic expectation that peace would usher in the dawn of an "Aerial Age" in which women would have a significant part. This image ultimately collided with the economic reality of diminished demand for the production of airplanes. Certain social expectations had a negative impact also. Definitions of postwar normalcy were based on prewar stereotypes. For example, as V-E Day drew nearer, in contrast to earlier advertisements that had exhorted women to work in the factory, the popular press, at the government's instigation, began to encourage women to return to the home and family. Home was where most of the WASPs, WACs, and WAVES went, though somewhat less eagerly than the women working in the aircraft factories.[1]

Those women who remained in aviation jobs began creating the legal framework necessary to reinforce and encourage continued change in American attitudes concerning gender roles for men and women. Society became accustomed to, although not fully accepting of, women in non-traditional occupations such as on the assembly line or in the military. Military women did not want to relinquish their newly gained positions. It was during the late 1940s that the military women's corps became permanent components of the services, partly as a result of the Women's Armed Services Act. The various women's aeronautical associations began to reorganize themselves after a wartime hiatus in activity. The aviation industry, after retrenching somewhat, began reorganizing for peacetime production.

It was the era of the nuclear bomb, the Marshall Plan, and the advent of Cold War relations between the United States and the Soviet Union. This climate of international affairs influenced all aspects of American life, from politics to technology. Aeronautical technology was intimately involved in these disparate spheres of activity, and it also became a positive factor in the American image and national pride. Certainly aeronautics represented the key factor in the nation's defense program. "The Bomb" may have been the ultimate weapon, but it was the airplane that delivered it.

At the time of demobilization, women in each of the different aviation areas were in the process of both challenging and adapting to the institutions that, in the past, had been organized and run primarily by men and which had frequently barred the participation of women altogether. The process involved a gradual but persistent questioning of the narrow expectations concerning the appropriate role of women. The resulting changes would set the pattern for subsequent generations.

Most women in the aircraft industry left the factory. Women were encouraged to do this for reasons other than the drop in production. A well-known Madison Avenue advertising firm, Batten, Barton, Durstine & Osborn, put out a pamphlet entitled "Women and Wrenches" explaining what they viewed as the chief effects of women's production-line labors.

Women may have acquired a passion for machinery, although laying a wrench on a certain nut, or spot-welding identical spots several hundred times a day, is hardly a comprehensive course in engineering, metallurgy, and inorganic chemistry. We venture the cautious prediction that when dungarees and lunch-boxes are put aside, the girls will continue to be more interested in what a thing will do for them then in how or why it works. However, one good may result from women's adventure in industry. They can hardly fail to appreciate the integrity of American production.[2]

The pamphlet described a postwar world filled with all sorts of technological wonders that would require considerable adjustment. However, the company noted that they thought it very unlikely that it would have to hire engineers and scientists to create advertising copy to "cater to machine-mad women."

Women working in factories did not continue to attract the media attention they had during the war. The jobs had lost the glamorous aura of wartime patriotism, and further, company executives could see no advantage in attracting women to the plants. It was generally expected in this postwar period that young men and women could finally resume their lives, establishing the households and families that had been delayed for several years.

The aircraft industry was thrust into a period of extreme instability as the Army Air Forces and Navy demobilized, and the government began selling surplus warplanes. The commercial airlines failed to flood the market with new orders. In addition to the drop in demand, the composition of the labor force was also being altered. During the war most workers were involved in production, but the ratio of engineers to line workers began to change in favor of the professional staff. This trend had particular influence on the employment of women.[3]

The AAF became thoroughly committed to a program of research and development for jet fighter and bomber aircraft. In all, 15 new programs were started, but this work had little immediate impact on production. The Navy ordered a few aircraft that had been developed late in the war, but on the whole, government business was quite small. This had been expected. What had not been anticipated was the sluggish nature of the commercial airline and general aviation segments of the business. The first year after the war had witnessed the opening of many new small airports and an increase in the numbers of students. This proved a short-lived phenomenon, however, as aircraft-operation costs overtook salaries and disposable incomes. The market for new planes bottomed out after a brief flash of intense activity, and hundreds of thousands of workers were laid off.

In October 1947, only 272,000 workers remained in the aircraft manufacturing industry, half the number employed in 1941. Interestingly, the percentage of women employees did not drop to 1941 levels. In October 1947, 28,500 women were employed in the industry, representing 11.8 percent of the total. Yet five years earlier they had composed, at 23,000, a mere five percent of the industry's work force. The 11.8 percent mark was as low as it would ever get after World War II, but it was typical of the demobilization period.[4]

The year 1948 was a very bad year for the aircraft industry. Ironically, its sales were nearly twice prewar levels, but a much greater overhead precipitated heavy losses. The industry eventually sank from its first place rank among American businesses in 1944 to 44th, based on dollar value of production output. It was at this point that the aircraft industry turned to the federal government for assistance—an action that would affect both the financial condition of the industry and the composition of the work force.

Initially, federal aid triggered a small rise in the number of women employed. Out of a total of 266,000 workers in September 1949, 12.5 percent, or 33,000, were women. They were limited almost exclusively to production workers or clerks, although a few women worked in the white-collar occupations of engineering or management.[5]

Betty Gillies was an example of a woman working in one of these white-collar positions. She was hired for a brief period to work as a test pilot for Ryan Aeronautical Corporation. This unusual opportunity was made possible by the fact that her husband was in charge of the test program for BT-13s and FR-1s.[6]

FIGURE 32.—Olive Ann Beech cofounded Beech Aircraft Corporation with her husband Walter in 1932. The only woman to head a major American aircraft company, she continued with the company in 1985 as chairman of the board. (S.I. photo 80-16080)

Olive Ann Beech, who had founded Beech Aircraft Company in 1932 with her husband Walter, was one of the rare women in aviation management. One of the chief executives of the company, Olive Ann Beech often promoted other women involved in aviation. She was behind the donations of Beech airplanes to Louise Thaden for the 1936 Bendix Race (won by Thaden) and to Ruth Nichols for her Relief Wings organization. Though Walter Beech retained the corporate presidency until his death in 1950, he became quite ill in the early 1940s, and his wife assumed all of the day-to-day work involved in running the corporation. She was responsible for preserving the company's economic viability throughout the postwar period, as well as for developing a program of diversification to extend services beyond the general aviation market. She also continued her interest in women's advancement and was credited with forming the liberal labor policies at Beech. These included special attention to situations affecting women in the training and personnel programs. She created a special center for women to increase their knowledge of business and industry. Her

FIGURES 33.—Flying jobs open to women were limited during the postwar period. Ex-WASPs were fortunate to be able to find jobs ferrying both new and surplus aircraft from one location to another. 33*a*, (left to right) Byrd Howell Granger, Ruth Dailey, and Delphine Bohn were three WASPs employed to ferry a shipment of Beech Bonanza aircraft to Brazil in 1947. 33*b*, Bohn is shown in Rio de Janeiro, Brazil, completing her delivery. (Courtesy of Delphine Bohn, 33*a*, S.I. photo 86-5581; 33*b*, S.I. photo 86-5579)

work was recognized (as happened many times later in her life) just one year after she formally assumed the Beech presidency, in 1951, when she was named "Woman of the Year in Aviation" by the Women's National Aeronautical Association.[7]

Other women combined technical talents with more traditional feminine roles. The postwar period marked the advent of the "flying secretary." Jeanne Coffey, who had been educated at Stephens College, worked for the Aeronca Company. She was a secretary, officially, but she spent her afternoons instructing potential customers, test flying new aircraft, or ferrying Aeronca airplanes from the factory to the buyers' local fields.[8] Elizabeth Gardner, who worked for Piper Aircraft before the war, then served as a WASP, returned to Piper in 1945. She worked in the public relations office and was often called on to use her flying skills. Soon she was writing all of William Piper's speeches; later, she was entrusted to deliver them to audiences in the Defense Department and to other companies involved in aviation.[9]

Gardner was not the only WASP who wanted to stay in the aviation business. Many of the former WASPs were desperately seeking jobs in the field. Most had to settle for roles peripheral to piloting, construction, or design, and only rarely did they fill supervisory positions. Former WASP Yvonne C. "Pat" Pateman worked at a California airport. She pumped gas, ran an airport cafe, and delivered airplanes. She became assistant manager after about a year. There were so many used aircraft for sale that Pateman had many opportunities to ferry small airplanes between the distributor and the local field.

In the spring of 1945, jobs ferrying surplus warplanes became available. The Defense Plant Corporation and Reconstruction Finance Corporation contractors had to supply their own pilots to fly the airplanes to the eight sales centers across the United States where the government sold its surplus warplanes.[10]

Delphine Bohn, a former WASP member and one of the "Original 27" WAFS, served as a ferry pilot with the Defense Plant Corporation before going to work at the Pacific Aircraft Sales Company. Pacific was a Beech dealership based in Los Angeles, which serviced California, Oregon, Washington, and Nevada. A "Girl Friday," as Bohn described her job, she served as office manager, secretary, test pilot, charter pilot, salesman, demonstration pilot, and ferry pilot. Bohn, who often demonstrated aircraft to famous actors and actresses from Hollywood, appeared to have a glamorous job. She worked very hard, but often her mink jacket and stylish clothing belied the high caliber of her experience and ability. In only a few years she was able to start her own aircraft dealership.[11]

FIGURE 34.—Delphine Bohn, former WASP and one of the "Original 27" WAFS members, in her office at Pacific Aircraft Sales Company in California. In 1946 she was a saleswoman/demonstrator pilot for this Beech distributorship, covering the California, Nevada, Oregon, and Washington territory. (Courtesy of Delphine Bohn, S.I. photo 86-5592)

Despite the diverse work of individual women in various phases of aviation's private sector, the largest group was still the flight attendants. These women, as well as their co-workers in ground occupations (like ticket sales and reservations), were caught in an unusual situation: an expanding job market offered tantalizing possibilities for raising their status and pay, but these opportunities were beyond their individual grasp.

Airlines executives shared the overly optimistic estimations of the aircraft manufacturers concerning postwar air travel. Both assumed that the civil aircraft market would replace the military demand, and they hoped to sustain their wartime revenue levels. Based on a doubling of passenger miles between 1945 and 1946 and the advent of transatlantic service, the companies placed hundreds of new orders in 1945 and 1946.[12]

As a result of the expansion of airline passenger service, increasing numbers of women were hired to work as flight attendants. The profession had become firmly established during the war, but now the women began collectively to desire an improvement in status, pay, and working conditions. One stimulus for this situation was that other airline personnel had successfully lobbied for such changes. Leadership of a movement to organize the flight attendants came in 1944 from Ada Brown of United Airlines. The task was a formidable one. Most women approved of the goals that Brown and others suggested, but the idea of a "union" was practically anathema. Unions, which then served mostly blue-collar laborers, were outside the experience of most of the flight attendants, whose upbringing was hostile to the overtones of radical activism from the 1920s and 1930s, which unions still had. Ultimately, they wanted to have their job viewed as a professional career and to call any association of flight attendants a "union" seemed to downgrade their status and job prestige. For this reason, Brown always referred publicly to the proposed organization as an "Association."

Appearance was still considered particularly important as was being unmarried. Instead of a nursing degree, applicants were now expected to have at least one year of college education. The job training had not changed as a result of the war, except that the women no longer had to concern themselves with wartime photography restrictions and baggage searches. For example, the course conducted by Transcontinental and Western Air (TWA) lasted six weeks and concentrated on "the principles of aerodynamics, aircraft meal service, baby care, first aid, and the social amenities."[13] In that era of piston-engined aircraft, the attendants had time to spend with passengers, and were expected to point out significant geographic features and answer questions about how the aircraft operated.

The women, who were required to be at least 21, were usually called "girls" and were treated as if they were minors. During training, they were housed in dormitory buildings under constant supervision of a headmistress (in loco parentis) who was the director of the training program and dormitory housemother. (This practice was also typical of many colleges in their treatment of their female students). The directors of many flight attendant divisions in the various airlines were women. While many of these women did not have great influence on airline policy, it is important not to underestimate their value as mentors to the younger women. They were competent, well-educated professional women who had decided against marriage in favor of their career. They served as vivid examples of the options women had apart from raising a family.

Most of the women they supervised chose not to follow their example. Even before the war was over, eight attendants a month, on average, were leaving TWA to marry. Yet apart from the year's service requirement initiated during the war, little attempt was made to retain their services. The reason for inaction stemmed from the management's conception of the job

and why women were suited to do it. Before the war, women were used to help convince passengers to fly, but as public air travel increased, the rationale changed. An article about Braniff flight attendants noted that

for certain positions it has been found that women are better adapted than men—for example, the air hostess. Since the early thirties, when the first young girl flew as the third member of the air crew, no substitute has been found for her position. Just as a woman is best suited to receive guests in her home, so she is employed by the airlines to make guests feel at home aloft in a plane.[14]

The name change from "stewardess" to "hostess" in the popular postwar press is significant because it placed the profession in the context of accepted roles for white, middle-class women. It was acceptable for women to work in this job because it was perceived as a short interlude of adventure granting polish, grace, and professional charm to bright young women. When people described the occupation it was always in terms of the benefits. Travel, shopping, and dining were seen as major aspects of the job description and not as perquisites of a demanding profession.

Some of the women in the profession had begun to see their work from a different perspective. Brown resigned her job as chief stewardess for United Airlines in December 1944, to become a full-time attendant again and to organize a union under the provisions of the Railway Labor Act of 1926. Her reason for doing this stemmed from her frustration at being unable to negotiate better wages and greater job protection while serving as chief stewardess. She began to organize the San Francisco region, while her colleagues Sally Thometz and Frances Hall assumed responsibility for the Denver and Chicago bases respectively.

United Airlines was very responsive to this work. William Patterson, president of United, believed that unions served important functions in industry and society. He also recognized that, since the flight attendants held, in effect, short-term positions, it would be much easier to work with a single established agent such as a union negotiation team rather than try to bargain with each individual. Within two months, 75 percent of the United attendants had signed "authorization-to-act" cards, and elections were held. Ada Brown became president; Frances Hall, vice president; Sally Watt, secretary; Edith Lauterbach, treasurer; and Sally Thometz, conferee. A constitution and bylaws were written, which officially established the group as the Air Line Stewardess Association (ALSA) on 22 August 1945. The women had also begun to prepare a contract to present to United.

The group initially was entirely within United, but its goal was to represent flight attendants on all airlines.

Hence, the women maintained close contact with the Air Line Pilot's Association (ALPA), which had a similarly broad-based membership. ALPA wanted to organize all of the airline workers but also wanted control, especially of any organization of flight attendants. For this reason, ALPA president, David Behncke, saw the autonomous ALSA as a threat. He was annoyed when ALSA failed to cooperate with his plans. Although ALSA had originally asked ALPA for advice in organizing, it was not at all interested in being subordinated to ALPA. Brown did not hesitate to tell Behncke this.

Without the support of the pilots, ALSA had to face the difficult task of negotiations without the resources of experience and money. ALSA leaders were always hampered by the turnover of membership, and constant vigilance was needed to keep the membership rolls current. Undaunted, Brown submitted ALSA's first request for conference dates in August 1945 in order to negotiate a new contract for the United flight attendants. That first meeting was a long struggle to find a common ground. The major point of discord between ALSA and United was over the reduction of the number of hours flown each month. United at first failed to take the women's demands seriously. Each group stood firm, and the situation was tense, with the company seeming almost to be avoiding progress.

The company lawyers eventually agreed to arrange meeting times around the flying schedules of the union team members (who could not afford not to work), and United always provided free transportation. In March 1946, after a seven-month effort, ALSA decided to consult the National Mediation Board. This resulted in a mediation process that lasted over a month. The time was well spent, for, in the end ALSA received a positive ruling. On 25 April 1946, Brown, Thometz, and Hall signed the first contractual agreement with United. In addition to gaining a contract, ALSA forced United to recognize the Association as a union—the first formal recognition of any flight attendant organization.[15]

That first contract served to legitimize the profession, increase wages, and improve working conditions. It also established a grievance procedure and legal protection from harassment and unfair labor practices. The starting salary was raised from $125 to $155 per month, the first such increase since 1930. A uniform allowance was provided, and mandatory rest periods were created. Flight time was limited to 85 hours per month with a maximum of 255 hours "within any three consecutive calendar months." Further, a stewardess could fly only eight hours a day and was entitled to a rest period of two hours for every flight hour.

The successive agreements of 1947, 1948, and 1949 were based on the 1946 contract. One unusual provision added in 1948 was a graded pay scale based on the type of aircraft flown. As the airlines adopted better equipment, flight times were being dramatically reduced. However, the aircraft inventory of the airlines was still quite varied, and thus attendants flying identical routes could end up with different flight times. The goal of the contract was to guarantee that attendants on the swifter airplanes would not be penalized with a reduced paycheck.

ALSA was beset with financial difficulties from the start. As a small union its greatest liability was its inability to administer and process the grievances of its membership. United was effectively free to violate contractual provisions because it knew that ALSA was financially unable to defend its interests. In 1947 the company reinvoked the requirement that attendants be registered nurses on its Hawaiian route. ALSA protested the move but failed to file a grievance officially until 1948. When the case finally reached arbitration, it was ruled that ALSA had followed improper procedures and could therefore no longer protest United's requirement.

Other challenges where brought to the union in 1947. Ada Brown married, thus becoming a "no marriage" rule casualty. Further, other groups were beginning to form all around ALSA to compete for the participation of flight attendants. In particular, ALPA had created a rival group known as the Air Line Stewards and Stewardesses Association (ALSSA) to compete directly with the "recalcitrant" ALSA. ALSSA, founded in August 1946, began a campaign to incorporate ALSA into the fold.

It was not until 2 December 1949 that ALSSA was successful in its attempt. The merger had been proposed by Irene Eastin, ALSA's third president. Following the unsuccessful protest of the reinstatement of the nursing requirement, it was readily apparent to her that, without a broader economic base and a larger membership, the union would fail altogether. The decision was made to join ALSSA in order to prevent losing all ground, but there were serious doubts. The United flight attendants were not convinced that ALPA, ALSSA's parent, really cared about them at all. Further, they did not want to lose their right of self-determination. Nonetheless, ALSA reluctantly agreed to come under ALPA's umbrella.[16]

Nonprofessional organizations also experienced much change during demobilization. The women's flying organizations spent this time getting themselves back into regular general aviation activities—especially flying. Like the flight attendants, middle-class women who wanted to fly had to "prove" that it would not affect their housekeeping abilities. In March 1945, *Woman's Home Companion* sent staff writer Mary Merryfield to

Parks Air College on what Merryfield described as "the not too flattering assumption, I suspect, that if I could [learn to fly], any of their readers could."[17]

Merryfield made several important observations about her experience. First, both the school and her fellow students (mostly men) expected she would fail because she displayed the "normal" feminine interests in home and husband. Second, she discovered that only by her determination to succeed could she demonstrate that these interests were not mutually exclusive:

I hope I proved to a lot of husbands who were in my flying class that we housewives can love our homes and still get up the momentum to fly a plane. My own husband for one, is pretty pleased that his wife can keep up with modern things and still darn his socks properly.[18]

Merryfield, in fact, was asked to talk to many of the wives of her fellow students in the hope that she could convince them to obtain their pilot's license. The men realized that it was not improbable for a woman to fly and that it might be enjoyable to share certain aspects of general aviation flying.

Many groups made formal attempts to convince women of their part in a postwar air age. The Women's National Aeronautical Association (WNAA) had been founded in 1928 to provide both a forum and an educational opportunity for women associated with men who were involved in aeronautics. During the postwar years, WNAA became more prominent, giving awards and even publishing a bimonthly magazine, *Skylady*. The WNAA national president, Gladys Dallimore, summarized the organization's goals as furthering aeronautical education of the public, stressing the safety of flying, and promoting the achievements of the airlines. When asked what the significance of the WNAA was, she responded,

What does all this mean—membership increased—students stimulated—women conscious of the air ocean—men gradually acknowledging that there is a place for women in aviation? . . . It is for the WNAA to do a good job, to fulfill its mission of putting forth a combined effort to promote, develop and stimulate the aviation industry, if wings are to have their rightful destiny.[19]

The WNAA tended to project a supporting role for women, rather than encouraging women to assume leadership roles such as pilot, business executive, or military officer. The WNAA defined itself as an organization for the wives, daughters, mothers, and sisters—any women associated with men in aeronautics—which would enhance their enthusiasm about aviation. As the notion of an "air age" gradually disappeared, the group, which had focused so closely on this faddish image, lost its purpose.

Changing times also overtook the two major wartime flying groups for women, the Ninety-Nines and the Women Flyers of America. The Ninety-Nines, alone, was able to survive, owing to its members' concentration on meaningful activities and issues centered around their interests as established pilots. The WFA eventually disbanded in 1954 in large part because they had concentrated on boosting opportunities for new people to take flying lessons at a time when the number of interested potential students was dwindling.

The postwar period was a turning point in the history of the Ninety-Nines. As a result of the WASP, CAP, and CPTP programs, a large number of new women pilots had been trained. For the first time it was possible for the Ninety-Nines to have a significant number of members who were professional, not avocational, flyers. Prior to the war, the club had always been defined in terms of the needs of recreational flying. The Ninety-Nines were proud of their members who had become WASPs, but despite debate on how best to expand the membership, genuine efforts to assimilate the wider focus of this pool of potential members failed to materialize. In general, the interest was in reinstituting the same purposes and activities they had supported before the war, such as general education programs, air shows, and air races. Air marking was the "public service" activity most frequently promoted. It was an easy activity to organize and implement with a volunteer labor force, it was not very expensive, and yet it was an invaluable aid to the general aviation pilot.

The organization was concerned about the increasing costs connected with flying small aircraft, but it did not really seek to exert any influence on fixed base operators to lower costs, nor did it inaugurate any scholarship program to aid less well-off individuals. There was a direct correlation between the expense of flying and the kind of membership attracted to the Ninety-Nines.[20]

The Ninety-Nines approach was to appeal to new members through personal recruiting and air shows. It was a grass-roots program that emphasized the fellowship aspects of the organization. Unconsciously, the Ninety-Nines opted not to become a professional or business organization. If WASPs and other women who had worked in aviation in a professional capacity wanted to join, they were heartily welcomed, but it was the new members who adapted to the social club format instead of the group changing to meet the needs of prospective members.

Under the leadership of new president Jeanette Lempke, who was elected immediately after the war, the focus of the group became the rejuvenation of the women's air races. In November 1946, the first discussions were held about a Ninety-Nine sponsored event at the Miami Air Races. Within six months plans were underway to participate in the National Air Races. During this time membership was expanding, from 700 members in September 1945 to just over 1,000 in 1946.[21]

In February 1947, the Texas chapter of the Ninety-Nines sponsored a women's aviation convention. It was held in San Antonio, and over 100 women flyers attended the two-day affair that included speeches by Hazel Raines of the Royal Air Force women's section, Edna Gardner Whyte, Blanche Noyes, and Dorothy Lemon. The event served to boost preparations for the first All-Women Air Show, which followed one month later in Tampa, Florida. An estimated 13,000 spectators watched the program of records, races, and aerobatic contests, encouraging the Florida Chapter to plan a second event one year later. The event was held to raise money for the Amelia Earhart Scholarship Fund (a program to help members acquire advanced ratings), and to promote public relations.[22] An editorial in the *99s Newsletter* stated:

The show will be presented in such a manner as to cause the Mrs. Average American Housewife to become more air-conscious and realize the importance of the aviation industry to our national economy and security and that they have a place in the air world.[23]

The event prompted air derbies and air-marking parties throughout the United States, but in Miami viewers were treated to parades, precision-flying competitions, sailplane demonstrations, parachute jumps, races for all categories of aircraft, and aerobatic flying.

A tradition of annual women's air races was begun. Organized by Mardo Crane, a former WASP, on behalf of the Ninety-Nines, the first race was held in 1947. That event, which ran 2,242 statute miles from Palm Springs, California, to Tampa, Florida, only had two entries, but the following year the field expanded as six aircraft and seven pilots competed for $1,500 in prize money. That race, like the one that followed in 1949, was called the Jacqueline Cochran All-Woman Trophy Race. (Cochran had graciously donated the prize monies.)

The 1949 race marked the formal beginning of the "All-Woman Transcontinental Air Race" (AWTAR). Crane, Dee Thurmond, and Irma Story of the Los Angeles chapter drafted the first real set of rules and regulations and developed an official timekeeping system (an honor system had been used previously), which firmly established the race as an annual event.[24]

These standards were based on the experiences of the second All-Women Air Show in 1948, which included that year's Cochran All-Woman Trophy Race. This show had resulted in a tremendous swell of pride

in accomplishment, a sense that, for the first time since before World War II, there was more to prove than flying skill. Blanche Noyes, one of the most prominent women in aviation, wrote this about the show:

At a national aviation meeting a year ago, I heard a woman make a speech about women fliers looking like bums at airports. I wish that woman could have been present to see these charming, well-groomed girls and women who were at the air show. Any one of them could have taken the prize for the best dressed woman in America.[25]

Noyes, who was the president of the Ninety-Nines from 1948 to 1949, had captured the essence of the unique experience of women pilots. This was maintaining a feminine appearance, planning activities independent of the major "male" shows and races, and always behaving like "ladies." In spite of the seemingly unfeminine activity of its membership, the Ninety-Nines chose to follow the social dictates of American society in determining its organizational activities and priorities. It was a women's club that *politely* broke with convention, an attitude that affected the diversity

FIGURE 35.—Blanche Noyes headed the Air Marking Division of the Civil Aeronautics Authority (CAA). Noyes served as president of the Ninety-Nines from 1948 to 1949 and later was a member of the Women's Advisory Committee on Aeronautics in the 1960s. (S.I. photo 86-12150)

of its membership, such that there were few members from minorities.

Postwar female flying activities were not limited to adult women. The Girl Scouts started a program known as Wing Scouts to encourage young girls to get involved in aviation. The heyday of this Scout group was from 1945 through the early 1950s, a time when young people were especially enthusiastic about flight. Exposed to the CAA's education series and excited by the exploits of wartime aviators, many children—both boys and girls—wanted to learn to fly. To enter the Wing Scouts, girls had to fulfill preliminary requirements, such as learning first aid and identifying at least 25 parts of an airplane. Once the prerequisites were accomplished, the girls were to specialize in one of four fields. Wing Scout "Cadets" had a scientific orientation and concentrated on learning the principles of aerodynamics. Wing Scout "Workers" focused on learning all about aviation careers with the airlines. Wing Scout "Builders" explored careers with aircraft manufacturers, studying production line techniques and learning about stitching, welding, inspecting, and other construction jobs. Finally, Wing Scout "Flyers" prepared themselves for the day when they would be able to become student pilots and learn to fly. Local chapters of the Ninety-Nines, the WFA, and the WNAA all helped sponsor Wing Scout troops. William Piper donated a Piper J-3 trainer craft to be used for giving the girls rides (no actual lessons were given).[26]

One activity that the Wing Scouts participated in was air marking. From its earliest days this endeavor had grown to become the cause célèbre of women in aviation. Women across the nation were painting signs to guide pilots, as part of a campaign for aerial safety and community service. Blanche Noyes was still in charge of that division for the CAA (and was the only woman who had permission to fly government airplanes). She never missed a chance to promote the cause of air marking. Now that the war was over and there seemed to be a surge in the number of general aviation pilots, Noyes believed it was critical not only to replace the signs removed during the war but also to add even more of these navigational aids. An active pilot and charter member of the Ninety-Nines, Noyes had instant rapport with all of the women's organizations. Her ability to win their support in this effort is an early illustration of the power of what might be called an "Old Girl" network: women pilots who had joined together in common interest and ability. No one else, male or female, seems to have ever tapped this network in the way Noyes did when she wanted to achieve a goal, such as her air-marking campaign, or have taken seriously the potential of this small and, therefore, supposedly insignificant band of women.[27]

Women in military aviation would eventually form similar networks within the ranks of their respective services. Until 1948, they were involved in the confusion of demobilization, and as members of small corps, the women were often situated in critical positions. Personnel officers and military leaders had the dilemma of wanting to release all the women immediately (the desire of each of the wartime women's corps' directors) and, at the same time, needing to keep many of the women in their wartime occupations to ensure the smooth functioning of a particular base or operation.

Only the Coast Guard SPARs were completely demobilized by June 1946 (the Coast Guard reverted to civilian control at the end of the war). The other services (the Air Force was soon established as a separate branch) found the dilemma complicated by internal debate over the question of whether women ought to have a role in the peacetime organization. Ironically, some of the very individuals who might have been expected to be most supportive—Oveta Culp Hobby, Mildred McAfee, Ruth Cheney Streeter, and Dorothy Stratton—all opposed the permanent retention of women. They never expressed their views publicly, however, and refused to testify before Congress in order to avoid the suggestion that they were criticizing the efforts made by women during the war.[28]

The military leadership ultimately decided that women should stay and they prepared to lobby Congress for authorizing legislation. The services would certainly have missed the female support staff, whose roles had become essential but were considered unsuitable for the returning male personnel. Each service tried a different method of handling the crises of personnel shortages. The Navy asked that at least 5,500 WAVES volunteers remain on active duty until July 1947. Further, the Navy had a drive to re-enlist 2,000 former WAVES in a newly created reserve group. Of the 14 major occupations available during this time, 9 were in aviation. There were approximately 1,500 enlisted women and 170 women officers serving naval aviation.[29]

The Army followed the Navy's example and also proposed a reserve organization. This proposal did not take into consideration that most of the WACs had already been demobilized. Also, the WAC leaders opposed the reserve plan, not because they felt there was no role for women in the Army, but rather because they believed the reserve plan as proposed could not be properly administered. In response to these objections, the Army decided to create a plan that would incorporate women directly into the regular service, instead of limiting women to only a reserve option.

The underlying question was still whether it was

FIGURE 36.—WACs train as Air Transport Command Stewardesses. Lifesaving instruction was an important part of the program to train the women to be able to handle emergency situations. As members of the air crew, the women were entitled to wear air crew wings on their uniforms. (NASM/USAF neg. 31295A.C.)

appropriate for women to become established professionally within the permanent military organization. Peace allowed the recurrence of the old nagging doubts as to feminine suitability in this tradition-bound male environment. Vocal and often virulent opposition to the presence of women in the military was voiced from many diverse sectors of American society. Individuals in positions of power, influence, and authority, however, did weigh in on the other side of this debate.

Eleanor Roosevelt was one of the most important advocates of women in aviation. Her insistence and

persistence in this area often proved to be the key to altering the debate from *whether* to *how* women could participate. Secretary of Defense James V. Forrestal, Generals Dwight D. Eisenhower and Omar N. Bradley (Army), General Hoyt S. Vandenberg and Brigadier General Dean C. Strother (Air Force), Admiral Louis E. Denfield and Vice Admiral Arthur W. Radford (Navy) all testified before the Armed Services Committees of Congress on behalf of legislation to keep women in the peacetime military. The view most effective in ultimately convincing Congress was not the philosophical

FIGURE 37.—Corporal "Torchy" West, WAC, was the only woman checked out as flight engineer on a transoceanic trip. Cpl. West was assigned to the West Coast Wing of the Air Transport Command and flew on Douglas C-54 aircraft. Here she is making the final preflight inspection of the airplane. (NASM/USAF neg. 31568A.C.)

argument on a woman's right to serve, but rather the shortage of "man-power."[30]

The Women's Armed Services Act of 1948 was passed on 2 June 1948, 206 to 133, and signed into law on 12 June by President Truman. Major General Jeanne Holm, USAF (ret.), described this event in her book on women in the military:

The act was many things to many people: to feminists, a leap forward for women's rights; to most women veterans, recognition of their contribution and vindication of their service; and to the military women who worked for its passage, sweet victory.

To the generals and admirals who actively supported it, the law was a vehicle for mobilizing women's skills in the event of a future national emergency and for meeting the military's more immediate requirements for volunteers, especially for those needed to fill clerical and other so-called women's jobs.[31]

Despite the occupational emphasis on clerks and typists, there were some opportunities for women in aeronautical fields. The Army dramatically curtailed the number of aviation specialities available to women

(this was the situation until the advent of women helicopter pilots in the 1970s). However, both the Navy and the Air Force did have many openings for women in aviation, though all were nonpilot positions.

On 7 July 1948, six new WAVES were sworn into regular service with the Navy. John Sullivan, Secretary of the Navy, remarked to those present at the ceremony:

Although you have been reduced to 2,000, it is my understanding that within the next two years you hope to induct 6,000 enlisted women into the WAVES, and I assure you the Navy will welcome them. In such highly specialized fields as communications, aerology, aviation training and many other fascinating and interesting fields they are better than any men we have ever had and we are proud to have them.[32]

In addition to the technical specialties, WAVES also worked with the Naval Air Transport Service. Women served as flight orderlies, the military counterpart to the civilian flight attendant. Other women served as Air Transport Officers (ATOs). The ATOs were responsible for all aircraft moving through their stations. They planned the loads, met the airplanes, and supervised

FIGURE 38.—Corporal Mary N. Dudley, WAC, was a parachute rigger and the base parachute-loft clerk at Offut Field, Fort Crock, Nebraska. (NASM/USAF neg. 34029A.C.)

the fueling, cargo and mail loading, and passenger boarding. They took care of all flight papers and were responsible for seeing that each flight was dispatched according to schedule. WAVES also continued to serve as tower operators, aerographers, and flight instructors (on training equipment only). Finally, Navy flight nurses and WAVES flight orderlies often worked together on air ambulance crews.[33]

One feature of the postwar WAVES program was the integration of black women. The WAVES had not permitted blacks to enlist until December 1944. They did not then establish any mechanisms for segregation since it was anticipated that very few black women would enlist, and thus it would not be necessary to expend administrative energy to separate them from the other women. At the war's end, there were 68 black enlisted women and six officers. In 1946 there were five black WAVES and one nurse. In February 1948, there were still only six black women in the Navy. Limited attempts had been made to recruit black women but to no avail. The Marine Corps Women's Reserve did not have any black women until March 1949. The Marine commandant was asked specifically if black women could join the corps. He replied, "if qualified for enlistment, Negro women will be accepted on the same basis as other applicants." The number of black women Marines was too small in this postwar period to find records of their job assignments or where they were stationed.[34]

The WAF (as Women in the Air Force were identified) was born with the Women's Armed Services Act, 12 June 1948, but it was essentially a reconstitution of the Air WAC. WAFs could occupy almost every occupation in the Air Force, from airborne radio operators on B-17 training flights to aircraft mechanics. The WAF did not have the WAC's rigidly prescribed structure and in theory the new program was intended to accomplish a full and complete integration of women into the Air Force. In practice, the WAF acronym helped to create an unofficial, but very real, distinct corps. Unfortunately, the "director" of this body had no power other than to serve as advisor, when consulted, on matters concerning women.[35]

By the end of June 1948, only 168 officers and 1,433 enlisted women formerly with the Air WACs had signed on with the WAF program. There were communication problems between the WAF and the demobilized Air WACs. Re-enlistment procedures were not clear and many women did not fully understand their options. These factors obviously contributed to the poor initial showing.[36] The goal of the WAF was to be an elite group of women modeled after, but superior to, the WAVES.

Because it wanted to open opportunities to women without college degrees, the Air Force did not have the same recruiting requirements as the Navy.[37] The recruiting results were disappointing: inadequate numbers were compounded by the poor quality of recruits that did enlist. Air Force leaders, especially General Hoyt Vandenberg, were surprised by this. Instead of choosing to work from within the service alongside director Geraldine May, Vandenberg would soon decide to try to solve the problem by seeking the advice of Jacqueline Cochran.

The five years following the end of World War II were exhilarating and exasperating for women in aviation. The "air age" previously anticipated by aviation enthusiasts seemed to have been illusory, yet American women were more air minded than ever before. Women were participating in nearly every phase of aviation. They continued even in the nontraditional fields of production and maintenance, where they had first gained entry during the war. Peace and a new economic environment produced more pressure for women to demonstrate that they were fully "feminine," an image that was, to many people, at odds with female involvement in aviation.

The various branches of the military were more progressive, at least on paper, than the civilian aviation world of the late 1940s. Regulations can only mandate standards of behavior; but the postwar rules for women in the military did mark the beginning of a change in American attitude. Women in military aviation offered valuable services and helped each of the different branches to meet personnel requirements. Still, the postwar period is characterized by the protracted debate over whether this was any place for a woman. The protests of enlisted men may have had some effect during this period, but they did not hold much ground once the Korean War began.

The small number of women and the lack of actual integration with men resulted in the women being grouped in small isolated corps structures. These women were convinced of the value of their service and excited by the unusual opportunities offered by military aviation, but they could not count on support from the public. It was an unusual and individualistic woman who chose a career in military aviation during the late 1940s, just as the women who joined the Ninety-Nines or continued to work in the plants were also unusual. At least by 1950, their corps, organizations, and networks were functioning well, so that those remarkable women were no longer alone in the aviation world.

6. "The Feminine Mystique" and Aviation: The 1950s

Have you tired of office job, or school, or nursing—
   Yet thrilled to the flash of silver in the skies;
If you can realize all aviations' problems
   But aid the vision of its future plans—
Then you will know this miracle of flying,
   The comradeship, the progress, and what's more—
You'll feel the very pulse beat of our country!
Welcome to the Stewardess Corps.

from "The Goal," Alma Frank

There were some remarkable women flying during the 1950s, women who broke out of "The Feminine Mystique," a concept that would soon be exploded by Betty Friedan's landmark book of the same title. They were not bound by the mythic glorification of housekeeping and child care in their quest for personal fulfillment and serious careers.

Most of the women in aviation were considered "unusual" in any circle, but there were some notable individuals who attempted specifically to extend the threshold of technological possibilities and perform new feats of adventure. Jackie Cochran crashed the sound barrier, Marion Hart flew the Atlantic (symbolizing a revival of challenging flights for light aircraft), and a major new organization for women in aviation was founded—the Whirly-Girls, the International Women Helicopter Pilots Association. Women, in small numbers, tenaciously hung on to positions in industry although its orientation had been transformed from the aircraft production-line to new aerospace ventures that required a dramatic increase in the numbers and roles of engineers. That change, of course, had major effects on the lives of women in military aviation, also. There was enough adventure and change, both technological and social, to ensure the vitality of most of the existing organizations for women in aviation. Clearly, the sustained presence of women in aviation during the 1950s was strong evidence of their collective influence and staying power, although their numbers were still small.

Among this group were women such as Ruth Butler. Growing up in Maine, Butler had learned to be an aviation mechanic during her high school years in the early 1940s. Her first flight, in 1944, was in a commerical airliner. Two years later, in 1946, she first put her hands on the controls of a Stinson. When asked, after soloing on 13 June 1952, what her ambition was, she replied: "To own an airplane and fly and service it myself and prove that there is a place for women in aviation. Nothing glamorous. Nothing sensational."[1] A year later she was part of an expedition to fly over both the North and South Poles (the first woman scheduled to do so) serving as a flight technician. Still, she claimed:

I don't want the glamour of stunt flying or racing or flying jets. I think it would be wonderful to be an airline pilot, but women haven't broken into the field yet. I think the ideal job for a women flyer would be as secretary-pilot to a business executive, flying his private plane and accompanying him on his trips.[2]

Other women proved to have as broad an imagination with regard to personal adventure and flying as Butler did. Marion Hart had always been a pioneering women professional. In 1913, she was the first woman to graduate as a chemical engineer from the Massachusetts Institute of Technology. Later she earned a graduate degree in geology from Columbia University. She sailed around the world in a 70-foot boat. For her, "distance and course" always represented the ultimate challenge. She did not learn to fly until 1946 when she was 54 years old. Eight years later, she enlarged the list of her many accomplishments by flying the Atlantic and breaking what she termed a "psychological barrier."[3] Yet, praised her friend Robert Buck (in a *Reader's Digest* feature, "The Most Unforgettable Character I've Met"), Marion Hart was not a strident feminist.

I've been amused more than once to watch her at an aviation gathering, where talk is rampant and male pilots tell tall tales.

70

FIGURE 40.—Jacqueline Cochran (left) and Jacqueline Auriol (right) were considered arch rivals during the 1950s. Auriol, who was a French test pilot, and Cochran traded the world speed record throughout the decade. (National Air and Space Museum)

FIGURE 39.—Marion Hart was a pioneer all her life. In 1954 at age 62 she flew her light plane across the Atlantic Ocean, just to test herself and her abilities as a pilot. (S.I. photo 86-12155)

She takes it all in and then, during a lull, will ask a question so quietly and humbly that a repeat may be requested. The question is invariably phrased to inflate the male ego; but if you know Marion you realize she already knows the answer—a lot better, frequently, than the man she asks.[4]

Jacqueline Cochran maintained her reputation as the most adventurous and influential of all women in aviation. During the 1950s, Cochran parlayed her considerable influence with the Air Force leadership into opportunities to break numerous aviation records. Eagerly, she was one of the first women to fly jet aircraft. Cochran did not want to set women's records; she worked to set absolute records. By June 1953, she held all but one of the principal world airplane speed records for straight-away and closed-course flight. Of particular note was her 20 May 1953 flight at Edwards Air Force Base, California, when she became the first woman to break the sound barrier, in a Canadian-

owned Canadair Sabre jet.[5]

These were the kinds of personal challenges that interested Cochran most. She preferred competing with men rather than participating in events such as the All-Woman Transcontinental Air Race. To Cochran there was no status associated with the women's events. A "woman's record" to her was second best. By competing with men she felt she earned the right to be part of the flying elite. She cherished her relationships with individuals such as Chuck Yeager, Fred Ascani, President Eisenhower, and the majority of Air Force generals. She had worked hard to be a part of their club and was usually associated with "women's groups or events" only in the role of benefactor. Her considerable wealth accorded her access to the highest echelons in politics, military affairs, and society, but her records brought her honors and public adulation. Among them was the Harmon Trophy, awarded annually to the outstanding woman aviator in the United States, which she won a number of times. These awards also extended her credibility as advisor on all things related to women in aviation. In many cases, however, her advice was detrimental to the existing programs and the participation of women. The worst example was her report on the women serving in the Air Force.[6]

During the fall of 1950, Cochran was requested by Air Force Chief of Staff, General Hoyt Vandenberg, to examine the WAF (Women in the Air Force). As a special consultant, she was asked to determine if the existing program represented the best and most effective use of women during peacetime. In the background of this question was the fact that America had become involved in the Korean War.

On 6 December 1950, Cochran submitted a report to General Vandenberg that railed against the existing program. Cochran opposed the concept of integration (of gender; race was not considered). If there had to be women in the military, and Cochran was ambivalent about that, then they must be the "best" women. The "best" was not defined according to the substantive issues of job performance. Cochran's report was extremely disparaging of the uniforms and physical appearance of the WAF. Cochran was right that, in the tradition-bound world of the armed services, if women wanted to become a part of the system, they were going to have to adopt all of the trappings of military docorum. Cochran's report, however, went further. According to Jeanne Holm, she equated good work with a particular standard of physical attractiveness, a criterion that must have caused offense even in 1950.[7]

Cochran certainly knew how to produce action among male Air Force leaders. She accurately sensed their attitudes toward women and capitalized on these biases, but it should be noted that at the same time she also engendered considerable animosity among the enlisted women and officers. They resented her ability, as they interpreted it, to "cavort" with generals and give advice on matters of which she really knew little. When the WAF was being established and the directors were working out the plethora of bureaucratic details, Cochran was traveling around the world enjoying privileges bestowed by General Arnold. According to WAF director, Colonel Geraldine May, Cochran had gambled away the future of the WASPs, and now it appeared she wanted to run the WAF. Her report proposed a civilian position of Special Assistant to Vandenberg; it was strikingly similar to her previous post as Director of the WASP.[8]

The Air Force was not willing to go along with Cochran's recommendation of a civilian corps director for fear it would stimulate other minority groups (in particular, blacks) to demand similar representation. Cochran's advice did result in a re-evaluation of the integration of women into the Air Force. She had proposed a separate women's group, and although Director Geraldine May stood firmly opposed to this, she could not match Cochran in her ability to influence

Air Force generals. The Air Force retrenched and made attempts to redesign its WAF program by imitating many aspects of the WAVES program.

May was replaced by Mary J. Shelley, who had to face the challenge of making the new program a functional one—despite the Air Force's inability to decide what it wanted. With the outbreak of the Korean War, the need for personnel increased, and the Air Force naturally wanted and needed to employ its female components. Air Force leaders also thought they could recruit women in the same manner and numbers as they had in World War II. They failed to realize that most Americans were not as eager to participate in this seemingly lesser conflict. Enthusiastic patriotism did not surface to help diminish society's anxieties about women in the military. The no-marriage rule was a great handicap; weddings became an easy way to break a service contract. These factors combined to make the recruiting goal—a leap from 8,200 to 50,000—not only unbelievable, but ridiculous. It was impossible to persuade that many women to volunteer (indeed, when the need to recruit an equivalent number of men had been discussed, the Air Force leaders stated that it would only be possible if there was a draft). The Air Force did not understand their recruiting failure and began to fault the women who were already in the program.[9]

The women who served offer a different perspective on the situation. Obviously, they wanted to be an integral part of the Air Force. To civilian women, WAF officers seemed to have made major advances in gaining equality. The key to success, however, was to be able to accomplish a military job while maintaining the external trappings of feminine behavior. This was a tricky balancing act, and it subjected women to the additional burden of scrutiny for reasons of their sex. For example, Lieutenant Colonel Marion Lee Watt, Chief of Classification and Evaluation Branch, Military Personnel Division, who was considered a successful WAF officer, received this complimentary assessment:

Tact, firmness, farsightedness and progressiveness mark her thought and action. It has been said, "She thinks like a man, yet is flatteringly feminine." What's more, she gets the job done.[10]

One thing that the Korean War did achieve was an end to the exclusion of women from nontraditional specialties. WAFs could be found in almost every aviation specialty, including airplane dispatchers, weather observers, blind-flying instructors, air traffic analysts, mechanics, and parachute riggers. Many Air Force women served as flight attendants. These women were full-fledged members of the Air Transport Service, although their jobs were similar to the civilian occupation. The difference lay in the fact that WAFs were part

FIGURE 41.—Yvonne Pateman (then a First Lieutenant) was the only woman stationed at Clark Air Base in the Philippine Islands in 1953. Pateman returned to the Air Force when the military "recalled" former WASPs to active duty during the Korean War. (Courtesy of Lt. Colonel Yvonne C. Pateman, USAF (Ret.), S.I. photo 85-15664)

of the military and had to submit to the discipline and rigor of the organization.[11]

The essential requirements for a woman's enlistment in the WAF were that she be 18 to 34 years old (women over 35 were eligible if they had served in the WAC during World War II and their number of years in service equaled or exceeded the number of years over age 35), a high school graduate, single, and have parental consent if under 21 years of age. To be eligible for Officer Candidate School, a woman had to be between $20^1/_2$ and $26^1/_2$ years old and have at least two years of college education, unless she already had enlisted status.

Basic training for both officers and enlisted personnel was held at Lackland Air Force Base in Texas and lasted between 8 and 11 weeks. The six major elements of basic training were general orientation, lessons in military correspondence, inspection and parade training, aircraft spotting, Air Force operations basics, and aptitude tests. After basic training, the women were sent to specialty schools for additional training.[12]

Utilization of women by the Air Force was not limited

to the WAF. Some civilian instructors and employees in the aviation field were also women. Gottie Schroeder taught in the Department of Engine and Technician Training at Sheppard Air Force Base. Her specialty was reciprocating engines, and she was qualified to work on anything from the 65 hp Continental to the Air Force's largest reciprocating engine, the Pratt and Whitney R-4360.[13]

Women had been in naval aviation for 10 years by 30 July 1952. They were working in almost every department at naval air stations and taught everything from the physiological aspects and use of ejection seats to parachute rigging. In 1952 about 13 percent of WAVES recruits were selected for airman training. As was true in the WAF (paralleling the role in civilian life), one of the coveted jobs was serving as a flight orderly with a transport squadron. Training and job assignments were, in general, integrated with the men. Other jobs for WAVES included air controlman, aerographer's mate, aviation electronics technician, aviation storekeepers, and trademan.[14]

Women Marines often worked side by side with

WAVES, and in one experiment three women Marines and one WAVE volunteered to try their hands at firefighting techniques used at airfields. The women who performed these tasks often had experience in other nontraditional jobs. For example, when the Navy lifted its ban on WAVE mechanics in late 1953, Ann Alger quickly abandoned her yeomen's job to join the flight line. Having tinkered with engines since the age of 14, she was a natural with airplanes, particularly the Corsair engines with which she worked most frequently.[15]

The only WAVE entitled to wear air-navigation wings was Commander Frances Biadosz. She served on the staff of the Air Navigation School during World War II, and after the war she became involved with the Navy's Public Relations Department. In 1959 she was assigned to NATO's Advisory Group for Aerospace Research and Development, where she served as a special assistant to the famous aerodynamicist Dr. Theodore Von Karman. Biadosz' attraction to aviation was shared by many other WAVES. Jacqueline Donnelly gave up her job as a New York telephone switchboard operator to become an aerial photographer for the Navy. "I have always enjoyed flying and being near planes," explained Donnelly, "and now I can do both. I've been on many assignments taking photographs and the training is exciting. I'll sign up again the instant my four year enlistment ends. I like being a WAVE."[16]

That attitude was produced by an environment of good organization. Joy Bright Hancock had continued in command of the WAVES program until her retirement in 1953. Her experience in two world wars and the demobilization period stood her well. The Navy leadership listened to her plans for recruitment, training, and deployment of women because they were consistent with the Navy's goals and philosophy. She knew how to work within the organization and make her presentations and proposals attractive to her superior officers (all male). Further, she had an understanding of women in the military that could be matched by few. For example, during the Korean War, she only proposed a 75 percent increase in female personnel, from 6,300 to 11,000 WAVES. Similarly, the Marine Corps only raised its ceiling from 2,250 to 3,000. When recruiting campaigns could only manage to acquire 8,000 WAVES and 2,400 Marines, the programs did not suffer the disastrous consequences that the over-zealous attitude toward recruitment and serious over-estimation produced in the WAF.[17]

A notable characteristic of women in the military was their ability at the administrative level to work together, regardless of branch affiliation. The directors often met with each other to discuss mutual problems and propose solutions. Another example of cooperation is seen in the joint training of Navy and Air Force flight nurses at Gunther Air Force Base, Alabama. The coordinated curriculum taught subjects like aeronautics, air evacuation, physics of atmosphere, and the effects of high altitude. Together they practiced water "ditching" (abandoning an aircraft over—or in—water) and other survival techniques.[18] This joint training program was a model the Department of Defense found useful, because it avoided the costly duplication of almost identical training courses. The development of other such joint service programs was part of the stimulus for the creation of a special citizen's advisory group, known as the Defense Advisory Committee on Women in the Services.

DACOWITS was the brain child of General George C. Marshall, then Secretary of Defense. In September 1951 he invited 50 women recognized as outstanding leaders in a variety of professions and civic enterprises to meet in Washington and advise on the subject of women in the Armed Forces. The problem of greatest urgency was recruitment, and to that end DACOWITS helped set up a unified recruiting campaign. DACOWITS became part of the defense establishment virtually overnight. By the end of the September ad hoc conference, a formal committee to work directly with the Assistant Secretary of Defense for Manpower, Personnel, and Reserve, had been created. Working groups were created in several areas: training and education; housing, welfare and recreation; utilization and career planning; health and nutrition; recruiting and public information; professional services (for nurses); and standards (recruitment).[19]

The group was a vital advocate for women in the services, but it also provided an essential and progressive forum for consideration of any question even tangentially related to women in military aviation. It consistently supported the right of women to participate in nontraditional fields. During the 1950s its concerns centered first on recruitment, and then on issues such as housing and public acceptance of military service and military careers for women. In 1955 members made the first of their many recommendations to the Air Force to include women in their Reserve Officer Training Program (ROTC), but it took until 1969 for the Air Force to comply. The committee was unique in the view it took of the direct connection between occupation and the quality of life enjoyed by an individual; from the start it asked the military to replace conventional definitions and sterotypes of women with a creative vision as to how to make the most of everyone's talents in service to the nation. Their advice was listened to and taken seriously (if not always acted upon immediately). Later the group would examine the really critical questions of women in aviation, including the training and use of female pilots.[20]

The military was not the only avenue for a young woman to enter the field of aviation. The Wing Scout program was still in existence, and several colleges were promoting aviation activities. Stephens College maintained its aviation studies program into the fifties, although the tone of its stated purpose had the clear ring of Cold War pressures: "Today the purpose of aviation studies at Stephens is to develop attitudes based on the realities of the Air Age since students will be participants in a society using airplanes as the dominating force for world unity." Many college women learned to fly at Stephens, and more than a thousand had completed the airline traffic course, when an article about the school appeared in *Flying*. One interesting statistic in this article is the fact that, in 1953, 90 percent of all students had parental permission to fly, whereas 10 years earlier, parents of less than a third of the women had responded favorably to the college's standard request for this approval.[21]

Other colleges had aviation programs for women. The National Intercollegiate Flying Association (NIFA) also sponsored clubs and flying contests. Jane Baker was a senior at Oklahoma Agricultural and Mechanical College when she won a major NIFA contest and was named NIFA College Woman Pilot of the Year. Baker had soloed in March 1953 and was a member not only of NIFA but also the Civil Air Patrol and the Ninety-Nines.[22]

The Civil Air Patrol had continued to grow after World War II to 91,000 members in 1956, 14,000 of whom were women or girls (teenage girls participated in the youth wing called the CAP Cadets). CAP pilots and observers flew more than 50 percent of all hours flown for aerial search and rescue missions in the United States. Harriet Robertus, the wife of an Air Force officer, was a typical CAP pilot. She first came into contact with the CAP when her husband became the CAP liaison officer for the Air Force. Always connected with airplanes (she worked for an Army Air Corps contract flying school and often flew with her husband), Robertus became an aerial observer for the CAP. When questioned about her priorities she responded that "next to my two children, Frances and William, the Civil Air Patrol comes first.[23]

Nona Quarles was also active in the CAP. She became a victim of instant notoriety when she decided to try for her pilot's license. As the wife of the Secretary of the Air Force, she had flown many miles as a passenger but had delayed taking flying lessons until 1957. Because of her husband's position, there were many obstacles to her ambition to fly jets. One article in *U.S. Lady* described the situation:

Her husband, Secretary Quarles, has interposed no objections to her becoming a pilot. He even indicated his willingness to be her passenger on occasion. But so far he has nixed her every chance to become jet borne.

She almost made it though, during the Air Force Convention in Washington last summer, Jacqueline Cochran invited her to crash the sound barrier [as her copilot] and Mrs. Quarles lost no time in collecting her flying gear and making her way out to Andrews Air Force Base. To her dismay someone had tipped off the Secretary, and Air Force officials at the field firmly but politely told her there would be no jet ride for her that day.[24]

Quarles wanted to participate in the CAP because of its service to the public. Accepting the fact that much of the media attention focused on her resulted from her husband's postion, she was able to use this situation to bring positive attention to the CAP. By way of reinforcement, she also demonstrated that her husband supported her aviation interests and activities.

Ruth Nichols was another famous female flyer with the CAP. A lieutenant colonel in the program, Nichols maintained a strong focus on youth, as typified by her frequent work with the CAP Cadet program. Apart from the CAP, Nichols was asked to make a global aviation tour to focus attention on the United Nations International Children's Emergency Fund (UNICEF). It was planned that she would be the copilot of the Douglas DC-4 on this trip but, unfortunately, she did not have enough hours of multi-engine flight time to qualify as a transport pilot on the DC-4. She did serve as a "courtesy 'extra' pilot," which allowed her to assist in operation of the craft. Her post-tour service to the CAP involved standardizing the Aeronautical Rescue Procedures of the CAP, including the services of air evacuation, air rescue, and air ambulances.[25]

The Ninety-Nines continued to be the principal women's aeronautical organization. In 1950 the Florida chapter sponsored their fourth and last All-Women Air Maneuver. The Amelia Earhart Scholarship Fund was a principal interest of the organization, but it aroused less zeal in the Ninety-Nines than the All-Woman Transcontinental Air Races (or Powder Puff Derby, as it was called by the press, following the example of Will Rogers, who gave the 1929 Women's Air Derby this name). Gill Robb Wilson, founder of the Civil Air Patrol and President of the Air Force Association, called it "the best single showcase of dependability of the light airplane and its equipment."[26]

In 1951 and 1952, in response to the Korean War, however, the race was called "Operation TAR" (Transcontinental Air Race) and was operated as a training mission to "provide stimulation as a refresher course in cross-country flying for women whose services as pilots might once again be needed by their country."[27] After

FIGURE 42.—Nancy Harkness Love, former WAFS director, stands with her three daughters in front of the family Beech Bonanza airplane on Martha's Vineyard in 1955. The Loves used the aircraft to commute from their island home to various points in Massachusetts and elsewhere in New England. (Courtesy of Margaret Love, S.I. photo 85-11129)

the war, the event became an official race open to all women pilots (not just the Ninety-Nines); it adopted knots and nautical miles as standard measurement and acquired a new leader, Betty Gillies.

Under Gillies' leadership the pilot requirements were stiffened. An FAI (Fédération Aéronautique Internationale) sporting license was needed, and all pilots were required to have a minimum of 100 hours of solo time, including 25 hours of cross-country flying. Further, a 10-year age limit was put on aircraft in order to allow for two-way radio communication and proper handicapping. The race was flown by a multitude of aircraft with vastly different airspeed capabilities. A "handicap," or par speed, was assigned to each aircraft in order to equalize the competition. The 10-year age limitation on aircraft eliminated highly modified World War II military planes from the event in order to assure the fairness of the competition. It was next to impossible to determine what the "handicap" of such a one-of-a-kind craft should be because there was no point of reference, such as other similar airplanes, by which to judge. Safety was always a major concern, and FAA inspections were required.

The AWTAR became a major event with its own office and a permanent executive secretary, former WASP Barbara Erickson London, who was hired in 1956. Run by a nine-woman board of directors, each AWTAR required a full year of preparation. The plans had to be coordinated with dozens of federal agencies, aviation associations, and other civic bodies.[28]

The race was becoming a cherished experience for its participants. The women eagerly donned coordinated outfits for their flights and post-race social events, but more importantly, they delighted in each others' company and the challenge of flying their aircraft over a difficult course.[29] Fran Bera was a regular AWTAR contestant. Flying in 19 consecutive races beginning in 1951, she placed in the top ten 17 times, with seven first place finishes to her credit. Her reflections on the race expressed a common sentiment held by participants:

I was very competitive and winning seemed everything. The AWTAR meant many other things to me, such as making friendships with other women pilots that have lasted a lifetime. These were women from all parts of our country and the world. They understood the pure joy of flight and the beauty of seeing our country unfold from coast to coast under our wings, while we tested our skills to the utmost. It meant the excitement of competing, while we took off into the rising sun on a still morning, and at day's end the fun and companionship we shared while telling our stories of frustrating minutes lost, the humorous things, the bad breaks, the good tailwinds and making it over the finish line just in the nick of time to miss the thunderstorm that would close the field. It was wonderful to communicate all of this to other women and know they understood.[30]

Over the decade, hundreds of thousands of miles were flown by contestants traversing the continent, as 499 aircraft and 905 pilots participated in this extravaganza of general aviation. General Hoyt Vandenberg,

FIGURE 43.—Fran Bera flew the All-Woman Transcontinental Air Race (AWTAR) for 19 consecutive years from 1951 to 1969 and then again in 1976 and 1977. Seventeen times she placed in the top ten, including seven first-place and five second-place awards. (S.I. photo 86-12153)

Chief of Staff for the Air Force, wrote to race chairwoman Mardo Crane, sending his regards to the 1952 contestants. He recalled the contribution of the WASPs during World War II as background to this forecast: "In flying the Sixth Annual All-Women Transcontinental Air Race the fact will once more be demonstrated that American women pilots can undertake exacting flights and complete them with safety and efficiency."[31]

Safety was always understood as a priority in the AWTAR, and gradually, after years of extraordinarily well-run events, the message could hardly have been made any clearer to the public—women were good pilots. Women were earning a collective reputation for safety consciousness within aviation circles that distinguished them from archetypical male pilots. They were known to avoid stupid and unnecessary risks; still, there was often a hint of sarcasm when people talked about women pilots, traceable to the popular image of a pilot as hero, bold adventurer, and daredevil that had emerged during aviation's golden years of the 1920s and 1930s.[32]

Women in aviation during the 1950s did not fit that image. Most women pilots held only a private license and flew recreationally. For the first time since the government began collecting aviation statistics, it stopped reporting separate totals for the number of male and female pilots (and for holders of various types of certificates as well). Thus it is difficult to document whether significant changes in women's participation in aviation occurred during the decade. Nonetheless, the anecdotal evidence is fairly abundant and suggests a steady, if moderate, growth in the number of women flyers. A good example of such anecdotes is to be found in Jo Eddleman's charming book, *Cows on the Runway*. Eddleman achieved a certain notoriety in 1957 when she promised to reward any of her junior high school math students with an airplane ride if they maintained a "95" average. Thirty-six students did, warranting a notice in Elsie Hix's "Strange as It Seems" nationally sydicated news feature.[33]

Jean Ross Howard broke another tradition, earning her helicopter rating in 1954, with the cooperation of the Bell Aircraft Corporation in Fort Worth, Texas. She had convinced company president Lawrence Bell that she ought to learn to fly a helicopter because it would enable her to do a better job as assistant to the director of the Helicopter Council at Aircraft Industries Association. Once she passed the CAA examination, Howard began to wonder exactly how many other women also had a helicopter rating. She discovered there were 12 others (excluding the possibility of some women pilots in the Soviet Union, who did not respond to inquiries), each of whom responded affirmatively to her suggestion that they form a club.

The Whirly-Girls, International Women Helicopter Pilots, came into being 28 April 1955. It was a very informal group at first, and Howard was chosen to serve as coordinator to handle correspondence and news announcements. The organization's initial goals were to "promote interest among all women in rotary wing craft, to establish scholarships to help other girls learn to fly helicopters and to provide a standby women's helicopter reserve for Civil Defense and other national emergencies."[34] Later, this last goal was replaced with the more broadly based aim of promoting heliports and landing facilities for hospitals.

The Whirly-Girls, like the helicopter, firmly established themselves in the aviation world, and by the close of the decade, membership had expanded to about 35. The helicopter played a role in the professional lives of many of the members. Dr. Dora J.

FIGURE 44.—Ann Shaw Carter was the first American woman to become licensed specifically as a helicopter pilot. She received her commerical license in 1947 and began working for New York's Metropolitan Aviation Corporation, flying sightseeing trips around Manhattan. (S.I. photo 86-12156)

Dougherty (later Strother) was a Whirly-Girl. A human-factors engineer with Bell Aircraft and a former WASP, whom Paul Tibbetts had trained to fly the B-29, she had been assigned by Bell to design helicopter cock-pits. Bell decided that, like Jean Howard, she too should know how to fly helicopters. Dougherty, a highly skilled fixed-wing pilot, found the rotorcraft not only challenging but also fun to fly. With only 34 hours of helicopter flight time, Dougherty set two world records for altitude (19,406 feet) and distance (straight line, 404.36 miles).[35]

Dr. Dougherty was in a profession that certainly could have used women. Aeronautical engineers were in great demand, as government contracts required ever-increasing levels of research and development. The industry was undergoing a major transition from aircraft production to aerospace engineering. The successful Soviet launch of the Sputnik satellite in 1957 marked a significant turning point. Only two years later, Aircraft Industries Association changed its name to Aerospace Industries Association (AIA). In its annual *Facts and Figures*, AIA noted:

78

Due to increasing pressures of the technological race in which the aerospace industry is involved, a wide range of measures are being taken by the industry to motivate, encourage, and in many cases finance young people of talent in pursuing higher education in engineering and the sciences . . . . Paradoxically along with the two-year recession in general aerospace manufacturing employment needs, there has been a continual recruiting plea for highly trained engineers and scientists.

The pleas of industry were obviously being heard, because between 1954 and 1957 the number of engineers and scientists rose 75 percent, from 48,500 to 84,900.[36]

Few qualified women could be found despite the demand. Industry did little to develop this "largest single source of new engineering talent," according to a U.S. Air Services article on women in aviation engineering. The problem was not entirely the fault of the companies, however. Estelle Elliot, an associate aircraft engineer with Lockheed Marietta, noted at a Wing Scout dinner that "most parents and teachers never think of suggesting engineering as a career for a girl, even though she may have shown mathematical, scientific or mechanical aptitudes." She noted that in 1955 only 10,000 engineers would graduate from college, yet there would be 30,000 new engineering jobs created that same year. To help dispel myths, Elliot told her audience of girls:

Let's face it. A certain amount of glamour exists because you crawl over, under and through an airplane and come out still being a woman! Many engineering jobs consist solely of desk work . . . . The course of study is not difficult, if you are mechanically or technically inclined . . . and if you are willing to work.[37]

Women did not enter aeronautical engineering except in very small numbers. At Boeing, for example, there were 47 women engineers employed in 1955, accounting for less than one percent of the company's engineering staff.[38] Many engineering schools did not admit women students. Those that did, did not actively recruit women. (It is interesting that of the few women who did major in engineering, most concentrated in aeronautics.) Women were not in general demonstrating any great eagerness to seize the opportunities in aeronautical engineering. College enrollments for women students were declining, and societal pressures dictated that the prime function of the collegiate experience was not education but matrimony. Further, engineering was perceived as a career; in the fifties men pursued careers, but most women did not.

Elaine Gething, a junior engineer in the aerodynamics section of Boeing's pilotless aircraft division, is an example of the rare female aeronautical engineer of the time. Gething had majored in mechanical engineering with a concentration in aeronautics at Oregon State University, but her first job, typical for most women,

even college graduates, was as a secretary in her home town. She hated it and quickly transferred to Boeing in Seattle, Washington, in 1950. Even so, an article describing her background stated: "One of these days she expects to marry and settle down to being a housewife, but in the meantime her work and year-'round schedule of outdoor recreation are a happy combination."[39] Gething's self-assessment illustrated the difference between men's and women's life and career expectations.

Women in significant numbers did continue to work in less technical positions in the industry, however. Throughout this period they averaged about 16 percent of the aerospace labor force, with the lowest percentage (12.4) occurring in September 1950 and the highest (18.0) in September 1952.[40] The fluctuations of the actual numbers of women employed followed those of the total employment curve. As production-line techniques came to require more sophisticated skill, women seemed to keep up with the trends. One benefit of working for this newly revitalized industry during the Sputnik era was good wages. Average weekly salaries in the aviation industry in 1959 were $107, based on hourly earnings of $2.68. This was about $5,500 annually, compared with a median female annual salary of just over $1,200; 45 percent of women working in 1959 earned less than $1,000 per year.[41]

The women in aviation who worked for the government were scattered in several occupations. Air traffic control was one of the more important fields. In 1959, 60 women were certified as controllers out of a total of 12,000. The Federal Aviation Administration sought to overcome its shortage of controllers by allowing trainees to substitute certain educational qualifications for the pilot's license requirement. Most of the women who entered the special training class in Oklahoma City had become involved with aviation during World War II. Male air traffic controllers were respectful of the women but always reminded them of the controller's guiding axiom: "You have to be right, or you don't work tomorrow."[42]

The business connections between government and industries were often made by women. This gave rise in the late 1950s to a group known as the Topside Aviation Club. Located in Washington, D.C., the club had about 50 members, all of whom were women in key positions in the aviation world. "Key position" was broadly construed to include women who were administrative assistants to top men in the airlines, government, the armed forces, and other aviation industries or associations. For example, group president Anne Meljunek was the administrative secretary for FAA administrator E.R. "Pete" Quesada. Essentially a social organization, the work of Topside Aviation Club members was described by *Air Force Times* as "often [smoothing] the way in lining up appointments for their bosses, coming up with hard information or merely 'oiling the machinery of mutual business.'"[43]

A world apart from the Washington aides in the Topside Club were the women who worked as aircraft mechanics. In the early 1950s, just before dissolving, the Women Flyers of America sponsored the first All-Women's Aircraft and Engine Mechanics class at the Teterboro School of Aeronautics in New Jersey. The course was open to 20 women and provided two scholarships.[44] Despite the fact that being a mechanic was an unusual job for females, women with sufficient aptitude and interest found ways to work on airplanes and engines all across the United States. In many cases women ran their own maintenance companies or aviation service businesses for pilots at small community airports.

Often escaping from those same small towns were women who became airline flight attendants. It was a job that one attendant, Lucille Chase, described as a combination of "hat check girl, nurse, babysitter, mother, cook, waitress, diplomat, psychiatrist, confidante and companion." The most important qualifications for an applicant were good physical appearance and the ability to get along with others. Whether a young woman underwent training at a company center or at a private school, classes taught inflight procedure, airline routes and codes, company history and policy (if a company school), stewardess regulations, and geography. Women also had to pass a CAA course in regulations and procedures (primarily safety procedures). Chase's descriptions in her book *Skirts Aloft* reveal that male chauvinism in the cockpits was a problem. Airline pilots and flight engineers often mistreated the attendants, making the women's jobs more difficult. She also noted that many passengers during this period were "first-time" passengers and consequently required a considerable amount of attention from the flight attendants.[45]

The attendants had been unified in 1949 under the Airline Pilots Association's Air Line Stewards and Stewardesses Association (ALSSA). One advantage of this merger was that ALPA performed the arduous work of bringing together the women on a majority of the commercial airlines. This had been one of the original goals of Ada Brown's ALSA. In 1953 there were 3,500 members, but the assets only amounted to $32,000—not enough to provide adequate services for members. The women, who constituted the vast majority of ALSSA's membership, still perceived themselves as an independent unit, and they attempted to function as such by electing officers and reorganizing their administrative offices. This naturally conflicted

FIGURE 45.—The most important function of a flight attendant in the 1950s was to present a particular image of youth, vitality, and feminine charm. As illustrated by this Frontier Airlines flight attendant, who is undergoing a "preflight check," matters of grooming were of prime importance. (Courtesy of the Association of Flight Attendants, S.I. photo 86-11868)

with ALPA's intentions to retain control over the flight attendant organization. ALPA had to contend with the women who were actively seeking their own charter with the AFL-CIO and both the Teamster's and Transport Workers unions, which had been expressing interest in having the flight attendants join their respective organizations. The antagonism between ALPA and ALSSA only intensified.[46]

During the decade, the advent of the jet had increased both the number of flights and the responsibilities of the flight attendants. The women wanted the 1959 contract to reduce the number of required monthly flight hours. ALPA was preoccupied with conflicts within its own ranks between pilots and flight engineers. ALPA wanted the flight attendants to drop their demands, claiming that the question of pilot salary was more important. It was an issue of health versus money, and the battle caused upheavals in union leadership and the loss of some attendants to the Transport Workers Union. It was the start of a struggle that would lead eventually to the re-creation of an independent flight attendants organization.[47]

The "air hostess," "stewardess," or "flight attendant," still retained her primacy as "woman in aviation," but female engineers, pilots, and executives were increasingly important. The end of the 1950s saw women in aviation on the threshold of a turbulent decade in the struggle for an even greater role. At the moment of transition, Gill Robb Wilson reflected on the presence of women aloft. His remarks are strikingly similar in tone to those of Hap Arnold's, spoken 15 years earlier to the last graduating class of WASPs.

The presence of women at the controls aloft, stunting in airshows, sometimes winning races in open competition, and spanning the continents and oceans in solo projects was a vast leavening in the dour opinion held by the public concerning the ultimate utility of flight.

And it must be remembered that the women did not easily come by their skills in the cockpits. There is always that element of pseudo masculinity that sustains its own ego by isolation of competition it cannot face. This the ladies met head on and overcame with the logic of tophand performance. In due course the bond of common experience became the criterion of pilot fraternity for men and women alike—and there it stands today. But it was a point that had to be settled if aviation was to be something more than a parade of muscles and myopia.[48]

Wilson's words were part reflection, part optimism. In the less prominent areas of aviation, such as light-plane flying and production-line assembly, greater equality had been achieved. The doors of power— the airline cockpit, the military flight schools, and the industry board room—were still closed to women. These were the places that women in aviation in the 1960s would attempt to integrate; they were ready for more than general aviation.

7. The Impact of the Women's Rights Movement: The 1960s

"An employer may not discriminate in his employment practices on the basis of sex."

Civil Rights Act of 1964

Toward the end of the 1950s, highly confidential experiments had been conducted by the Air Force on the physiological characteristics and suitability of women for space flight. Ruth Nichols, founder of Relief Wings, had participated in a set of these "astronaut" tests at Wright-Patterson Air Force Base in Dayton, Ohio, during which time she suggested that aerospace medical researchers ought to compile data on women as well as men.[1]

Then in September 1959, Geraldine "Jerrie" Cobb, a young talented pilot, was introduced to Dr. W. Randolph Lovelace of the National Aeronautics and Space Administration (NASA) and Brigadier General Donald D. Flickinger of the Air Force, both distinguished in the field of aerospace medicine. Cobb was 28 years old, with 7,000 hours of flight time, three world records, and the FAI (Fédération Aéronautique Internationale) gold wings of achievement. She was a pilot and manager for Aero Design and Engineering Company, which manufactured the Aero Commander aircraft, and was one of the few women executives in aviation. She was willing to give up this career in order to become the first woman to undergo the Mercury astronaut tests at the Lovelace Foundation in New Mexico.[2]

Cobb successfully completed all three stages of the grueling physical and psychological tests that were used to select the original seven Mercury astronauts. Her performance led to further tests for women—this time a group of twenty, all of whom were selected because they were outstanding aviators. Cobb's scores also led to the proposal by some NASA officials and military officers that the United States be the first nation to put a woman in space. Twelve women passed the first two rounds of tests and were scheduled to undergo the final round at Pensacola Naval Air Station in Florida. NASA refused to authorize the completion of the tests for fear that such action might be taken as approval of female astronauts. The decision to cancel the tests was made just the day before the women were due to leave for Pensacola, causing them to feel considerable disappointment and anger toward the NASA administrators. The resulting controversy involving NASA, Congress, and a handful of determined young women set the stage for a decade of furious debate on the role of women in aviation.[3]

Jacqueline Cochran was once again in the fray. Cochran paid all of the test program expenses for the women "astronauts," but then testified during special Congressional hearings against the idea of including them in the space program. She believed that the only way to include women was through a trained corps of specialists, strongly reminiscent of the WASPs and WAFs, but, she stated, the nation could not afford the time or money to conduct such a program. She spoke of women's lack of commitment and high attrition rates due to marriage and pregnancy.[4] In a personal letter to Cobb she wrote:

Women for one reason or another have always come into each phase of aviation a little behind their brothers. They should, I believe, accept this delay and not get into the hair of the public authority about it. Their time will come and pushing too hard just now could possibly retard rather than speed that date.[5]

This letter, which was secretly circulated among top Air Force and NASA officials, earned Cochran praise for her "statesmanlike" understanding of the situation.[6] Cobb and Jane Hart, wife of Michigan Senator Philip Hart and one of the successful test candidates, refused to take Cochran's advice and, instead, insisted on their right to participate. They lobbied Congress, Vice President Lyndon Johnson, and NASA administrator James Webb for inclusion of women in the astronaut program. NASA reacted by announcing a new formal requirement that a candidate must have experience as a jet test pilot. Prior to this time all of the candidates selected had been experimental jet test pilots, but it was not a stated requirement. This development negated the chances of selection for almost every woman in the United States—except Cochran, who, as a test pilot for two manufactuers of jet aircraft, had continued to remain visible in the aviation world by setting new speed records, such as the 100-km

FIGURE 46.—Geraldine "Jerrie" Cobb was a well-known pilot in the 1950s and early 1960s. She set several world records in Aero Commander airplanes and was the first woman officially tested by NASA as a possible candidate for the astronaut program in 1960. (S.I. photo 79-6359)

closed-circuit record on 1 May 1963 in a Lockheed TF-104G Starfighter and the speed record over the 15/25-km course on 11 May 1964, also in a Lockheed TF-104G. Without inside connections with civilian contractors that built the planes, the only way an applicant could acquire the requisite jet test-piloting experience was through affiliation with the military—and women were not allowed to fly for the military.[7]

When the Soviet Union launched Valentina Tereshkova into space in 1963, Claire Booth Luce wrote an editorial in *Life* magazine condemning NASA officials: "Why did the Soviet Union launch a woman cosmonaut into space? Failure of American men to give the right answer to this question may yet prove the costliest Cold War blunder."[8] Of course, the motivation of the space program was public relations, Luce argued, and NASA was being hypocritical to deny that propaganda and public relations were important components of the space program. Luce's comments were indicative of a larger struggle that was occurring in American society.

The question of civil rights had been much in the forefront of American consciousness for some time. Two laws were passed in quick succession that would prove to be particularly important to women in all professions, not just aviation. The first was the Equal Pay Act of 1963, which required equal pay for equal work. The second important law was Title VII of the 1964 Civil Rights Act. Title VII prohibited all discrimination on the basis of sex for any reason in determining employee compensation.[9] The laws empowered the government and individuals to take legal action against discrimination; and yet no significant change seemed to happen. The President's Commission on the Status of Women produced a lengthy report in 1965 that once again detailed wage discrimination and the declining number of women in professional and executive jobs but failed to offer any genuine prescriptions for change.[10] In June 1965, the National Organization for Women (NOW) was created. It became, among other things, an important advocate for women's rights in the cockpit.

Women in the aerospace industry found that the debate about women's rights initially had little effect on them. Most of these women were clustered in specific unskilled occupations and within these fields, "equal pay" generally was the norm even if equal advancement or career potential was not. One interesting note is that in 1962, Aerospace Industries Association stopped reporting the employment figures for women in aerospace. Some researchers, such as Cynthia Enloe, argue that this was a continuation of the sociological trend toward a diminished and less visible female presence in the aerospace industry.[11] The actual number of female employees, however, remained relatively stable. Women represented between 14 and 17 percent of the aerospace industry workforce throughout the decade. Fluctuations occurred as production responded to escalation of the conflict in Southeast Asia.

The disappearance of the statistics does indicate two facts. First is the lack of attention by the industry to the

FIGURE 47.—Dr. Jeannette Piccard was a scientist doing research on balloon flight and aerostatics beginning in the 1930s. Dr. Piccard is shown inspecting a mock-up of a Command Module. She became a consultant to the Manned Spacecraft Center at NASA's Houston, Texas, center during the 1960s. (Courtesy of NASA, S-65-24927)

contributions of women. This is not surprising, as the aerospace community had almost completely switched its emphasis to research, science, and engineering and away from production, where its women employees were concentrated. It was the leading employer in manufacturing in the United States, yet there was little status or prestige attached to the production end of the business, where no one, male or female, was receiving much recognition.

The second factor is the relatively small number of women in the aerospace engineering field. The lack of encouragement for young girls to take mathematics and science courses in the 1950s militated against the presence of women engineers in the 1960s. Only one percent of the doctoral degrees awarded in engineering were earned by women during this decade. Likewise, women represented about the same percentage (one percent) of the undergraduate enrollment in engineering.[12]

Nancy Fitzroy was one of the few women in the 1960s engaged in an engineering career. She began working as a heat-transfer engineer for General Electric in 1953. During the 1960s she worked on the thermal design of the afterburners for General Electric's J79 jet engine. To her knowledge there were only four or five other women engineers working for General Electric at that time, and while they did not seek each other out, she felt there was a certain unspoken bond born out of the common experience of being female in a male-dominated environment.[13]

Fitzroy was a pilot with both fixed-wing and helicopter ratings. Her work on a multitude of General Electric projects at various work sites provided her with the opportunity to commute by air (using her own airplane). This provided her the added benefits of increasing her piloting proficiency as well as cutting travel times to a fraction of what ground transportation would require.

In the late 1960s, she wrote a short pamphlet on her job as an engineer. It was part of a career-education series and contained information on her job and her lifestyle. In it she noted some of the difficulties she had when she was first hired. It took a major effort to convince the company that she was a serious career woman. The company's contrary stereotype resulted in her being passed over for job promotions or being paid a slightly lower salary than her male colleagues. The 1960s civil rights actions seemed to have lessened some of these discriminatory practices, but they did not succeed in removing them altogether.[14]

Fitzroy was a member of the Society of Women Engineers, which had been founded in 1952. The 1960s were a lean period for the society's membership; numbers were so small that it was difficult for them to be effective in realizing common aims—including their desire to see a greater number of female engineers. Women in aerospace engineering are rarely discussed for the decade of the 1960s because there were so few of them.

Another area that was hardly even considered was the hiring of women pilots by commercial airlines. Women did work as professional pilots: small airports, local flight schools, and light plane dealerships all hired women to fly for them. Jan and Marion Dietrich were two of these professional pilots; they were twins who had qualified in the Mercury astronaut tests. Jan was an FAA flight examiner, and Marion an aviation writer. Jan Dietrich described her reaction to the first time she was hired to work as a pilot (she gave rides to potential customers of an airplane dealer at a small airport): "Commercial women pilots were about as acceptable as female ship captains. I had never heard of anyone deliberately wanting to hire a woman pilot. This seemed amazing if not downright charitable."[15] Slowly the real reason for her selection would become apparent:

I demonstrated the planes, working hard making the tires whisper onto the runway. A passenger remarked, "Why the airplane practically lands itself," and Larry [her boss] beamed at me. Like breaking through the overcast, suddenly I realized the reason for having a lady pilot. If a girl does something well—especially a small one who dresses and looks like a girl—it seems easy, as if anyone could do it. And I hurried off to change the stocking that had just run.[16]

It was the old ploy of using women's apparent limitations to sell an aviation product, but change was on the horizon. In the 1960s, people were becoming considerably more aware of the hollowness of stereotypes associated with specific occupations. For example, an article on aviation careers described the desirable characteristics for an airline pilot:

The preferred applicant is married (this gives some assurance of stability and motivation) and although no one will say so, has completed his military obligations. He has a good achievement record in college with some interest in athletics and has a stable work history. In general he is an extrovert and has assumed leadership roles in his activities. He is outgoing, stable, mature, "outdoorsy," dominant (but not cocky) and shows judgment.[17]

Soon after this description appeared in *Flying* magazine, however, there were numerous objections from readers about the blatantly biased language. These protests demonstrated the slowly evolving trend of the time toward less gender-specific terms.

Spurred on by debates and public consciousness, highly visible organizations such as the aviation magazines were hiring women as writers and editors. *Flying* magazine inaugurated a regular column on women called "Skirts Flying." Month after month, author Sally Buegeleisen, herself a pilot, followed the progress of women in aviation. Through her articles she developed an open debate and public understanding of the subject of women in aviation that had never really existed before. From this vantage point she could see the connections between the women's rights movement and women flyers. Though not herself a radical, she could not help posing the questions that encouraged the male-dominated aviation community to think differently about women.

This was one reason why *Flying* encouraged its new editor, Patricia Demarest, to learn to fly helicopters. The magazine wanted to make her a successful editor, and at an aviation publication knowing how to fly was practically a professional credential. Without it, Demarest could hardly hope to be taken seriously and could expect to encounter barriers to her career advancement. Demarest had been an airplane enthusiast since childhood, but she had never taken any flying lessons. She described the trials and tribulations of learning to fly a helicopter in an article that appeared in *Flying*'s special August 1965 "Women in Aviation" issue.

Demarest made an important observation on the differences between men and women student pilots: "It's probably true that it takes a woman longer to learn to fly than it does a man. While a male student pilot is reviewing his last lesson and imagining his next, his student pilot wife is more likely to be cooking dinner and deciding if the ironing can wait until Wednesday."[18]

Demarest had identified an important social phenomenon. She and other contemporary journalists contributed to the late 1960s sharpening of the parochial debate within the aerospace world on the role of women in aviation. Robert Parke wrote an editorial stating forthrightly that

of course the presence of hordes of women in an activity does revolutionize the activity .... The hardy airplane salesman has known since the dawn of aviation that airplane sales would truly zoom—not if prices were cut, not if airplanes were prettier, not if the National Safety Council endorsed the airplane—but if women took to them.

To some the feminization of flying will be akin to holding a woman's bridge party in a monastery. And it's true the day of rowdiness of the Quiet Birdmen and the all-night hanger session may pass. But for many men this is a small price to pay when the approval of a spouse can mean the difference between flying and not flying.

And for those to whom flying is an escape from the cares of the ground—something to be savored best when solo—the beauty of the sky is that it is big enough for both of us.[19]

Evidence of the change in attitude toward women in aviation came in the quantity and scorn of the letters to the editor in response to Milton Horowitz's article, "For Men Only?" Horowitz claimed it was impossible for a real woman to fly because it violated her normal feminine sensibilities. Flying was the last bastion of masculinity, and the presence of women diluted the experience, according to Horowitz. His remarks provoked a wave of protest from both men and women, who totally disagreed with the sweeping generalizations made in the article.[20] "Why are men so afraid of flying women? Don't you know we're out to join you, not beat you?" wrote one woman in a letter to the editors of *Flying* in November 1965. That was precisely the message newly organized feminists were preaching.

One group in the aviation industry that was dramatically and immediately effected by Title VII of the Civil Rights Act was the flight attendants. The saying among airline executives had been, "Use them 'til their smiles wear out; then get a new bunch."[21] The airlines seemed oblivious to the discrimination involved in requiring attendants to resign when they married or reached the age limit of 32 to 35, depending on the airline. The airlines claimed that the "hostess" was the personal representative of the company. They felt they had the right to project an image of youth, freshness, eagerness of service, and seeming tirelessness. The same airline executives complained about the expense of having to train so many women. They neglected to mention that training was significantly less costly than funding pay raises, pension plans, and health insurance. They never discussed the airlines' fear of a flight attendant who was married and a part of a two-income household. Such a woman would be in an extremely advantageous position in contract negotiations because she could afford to go on strike. Hence, a powerful reason for wanting to continue the "no-marriage/out by 32" rule was the fear that these women would exert considerable influence over the company management and alter the status quo.[22]

These fears surfaced in the first suits brought against the airlines. In *Cooper v. Delta Airlines, Inc.* the court ruled against Eulalie E. Cooper, who had filed the 1967 case protesting that her termination with Delta was discrimination on the basis of sex. Delta only employed female stewardesses, and thus the court ruled the discrimination was between married and unmarried women, not between men and women. The case prompted the new Equal Employment Opportunity Commission to examine the question of whether discrimination based on marital status, when applied only to one sex, could be considered sex-based discrimination. Both the attendants and the airlines were eager to have a ruling.[23]

NOW played an important part in the hearings, actively supporting the attendants by helping with legal assistance, garnering legislatitive support, and drawing media attention to the issues being raised by the union. NOW's president, Betty Friedan, even testified at one set of proceedings.[24] The commission issued its opinion that marital restrictions did in fact violate Title VII. The opinion was nonbinding, but most airlines began to change their policies anyway. United Airlines, the largest carrier with the greatest number of flight attendants, refused to change the rule. Finally in 1968, United, still the last holdout on "no marriage," agreed to discuss the issue with ALPA and its Stewards

and Stewardess Division (which had replaced ALSSA in 1960). The union wanted to win reinstatement and back pay for flight attendants who, in their opinion, had been unjustly fired. United resisted each of the many proposals, although the company was financially capable of making such a restitution, though obviously at the loss of some profit. In 1969, a limited and marginally acceptable agreement was reached on reinstatement, but the question of back pay would not be resolved until the early 1970s[25]

The women who worked as flight attendants during the 1960s were still young, attractive, and mostly single. Salaries had improved, with base pay starting at $345 per month. Wages had built-in incremental increases for the subsequent 10 years of employment. Thus, top pay for an attendant with 10 years of experience would be $520 per month. Another issue of great import concerned flight attendants. Most of them were white, college-educated, middle-class women. Asian, Native American, and Hispanic women had been hired in limited numbers and primarily to service special routes in areas such as Hawaii, Latin America, or the Southwest. Virtually no black women were hired until the Civil Rights Act of 1964 was passed. Prejudice continued to operate against any large influx of minority women. Further, there is no evidence that the union leaders made any significant effort to alter the situation by encouraging the companies and their membership to increase minority representation. Nevertheless, the legal protection of the Civil Rights Act made it much harder to willfully exclude blacks and other minorities.[26]

The vast majority of the women pilots in the 1960s were involved in general aviation. Their numbers more than doubled from 1960, when there were 12,471 licensed women pilots in the United States (3.6 percent of 348,062 total), to the end of the decade, when there were nearly 30,000 women pilots (still only 4.3 percent of the total number of aviators, which had grown to 683,097).[27]

There were many organizations in the 1960s that focused on women in aviation, but such attention as the general public gave to aviation was usually drawn to remarkable individuals such as Geraldine "Jerrie" Mock and Joan Merriam Smith. Both of these women flew around the world in 1964, unintentional rivals in this, the first (and finally successful) two attempts of the feat by a woman since the disappearance of Amelia Earhart in 1937. Smith's route was a recreation of Earhart's famous, but ill-fated flight.

I had had the dream for years, first to fly an airplane, then to fly one as she [Earhart] did. When I was in high school, I would tell my friends and classmates that someday I was going to fly around the world just like Amelia Earhart. Everybody just laughed. They

FIGURE 48.—Flight attendants for Frontier Airlines pose for a corporate public relations photograph at Denver's Stapleton Airport in the late 1960s. (Courtesy of the Association of Flight Attendants, S.I. photo 86-11873)

knew I was a baseball-playing tomboy, and this was a tomboy fantasy. But I knew that since Amelia disappeared in 1937, no other woman had attempted to fly around the world. This only heightened my ambition to be the first one.[28]

Smith realized her dream, but she was not the first woman to fly around the world. Jerrie Mock in a single-engine Cessna 180 called the *Spirit of Columbus* earned that honor. Mock had applied for and received official sanction from the National Aeronautic Association (NAA) for a world record before Smith did. Because only one attempt at a time on this sort of world record can be made, Joan Smith's belated application for a virtually simultaneous try cost her the right of official

sanction. She decided to make the flight anyway, and since both she and Mock were departing about the same time during the spring of 1964, the press immediately presupposed a "race." It was hardly a race, because the two aircraft and the routes were very different, but the situation engendered a bit of competitive spirit and drew public attention to the participation of women in aviation.[29]

The Ninety-Nines were, of course, very interested in these flights. They also helped Ann Pellegreno with her round-the-world flight in 1967, a 30th anniversary celebration of Earhart's attempt. However, the prime interest and major commitment of the Ninety-Nines

FIGURE 49.—Geraldine "Jerrie" Mock became in 1964 the first woman to fly around the world. This photo was taken just prior to her March takeoff. (S.I. photo 86-12151)

during the 1960s was air racing. In addition to the AWTAR, they embraced the All Women's International Air Race, or "Angel Derby," as it came to be known. The race was open to any woman pilot, but the Ninety-Nines provided help in organization and management, not to mention forming the largest core of enthusiastic contestants.[30]

Doubts about the value of the events, especially the big race, the AWTAR, were smoldering in the minds of some observers. Sally Buegeleisen wrote:

There is also a growing skepticism among the racers about whether anyone with limited income has any chance of winning. "The present system," one of the women said, "rules out any possibility that any but a select few will ever be on top in this race." There is a growing resignation among the women who are racing that "people with average incomes . . . cannot afford to have their planes prepared" and so don't stand a chance.[31]

Buegeleisen's criticism may have reflected one concern of participants but it failed to convey one of the most distinctive features of this "race." People external to the event assumed that competition was the sole purpose of the race. There were occasional suggestions that all of the work needed to put on this event was not being adequately compensated or recognized. Recognition, to such observers, was usually understood in terms of male sporting events, where column inches on a sports page, sponsorship by aviation corporations, or a large purse was perceived as the standard. The

FIGURE 50.—The Whirly-Girls celebrated their tenth anniversary in May 1965 on the steps of the U.S. Capitol. (Courtesy of Jean Ross Howard, S.I. photo 86-12201)

intangible rewards, however, were of greatest value to the organizers and participants in this event. Pride was a big factor. The event, like the Ninety-Nines in many ways, was a self-contained entity and the participants did not measure success in column inches. Although public recognition through the media was always viewed positively, these women valued chiefly the fact that every year there was a celebration of women in aviation, and anyone connected with flight was a winner in this event.

Women were attempting to share the values of the race with the larger aviation community. The Ninety-Nines, like its sister organization, the Whirly-Girls (which had expanded to 140 members in 1969), were exploring different ways of communication. They first turned to their men's auxiliaries. By encouraging the presence of men at their functions, the women pilots hoped to open a positive dialogue about the role of women, both as individuals and as a group, in aviation. They wanted it to be taken for granted that they were serious and accomplished aviators.

The organizations however, were not all talk and philosophy. They were also interested in making real contributions to aviation, and they felt no compunction about seizing the appropriate moment. That is what the Whirly-Girls did, when at a tea with President

Kennedy their representatives actively lobbied for an increase in the number of heliports in metropolitan areas, especially Washington.[32]

One of the most promising means of communication was created by the federal government. On 4 May 1964, President Lyndon Johnson announced the formation of the Women's Advisory Committee on Aviation (WACOA) within the Federal Aviation Administration. The idea for such a group had been proposed by FAA Administrator Najeeb Halaby, and its purpose was to provide recommendations about public relations and education programs on flying as well as to encourage greater family participation in aviation. There were 32 members, 27 public appointees and 5 from government (serving ex officio), all women with outstanding reputations in aviation and community service. Of the original 32, all but two were or had been active pilots. The two who were not were the Director of Stewardess Services at American Airlines, the top female executive in that company, and a contract manager at Hiller Helicopter.[33]

WACOA was organized along the same lines as the Defense Advisory Committee on Women in the Services (DACOWITS) had been, and in fact met with their military counterpart during their preliminary planning meetings. Thus, they divided into five subcommittees,

FIGURE 51.—The Women's Advisory Committee on Aviation was founded on 4 May 1964 to serve as an advisory group to the FAA administrator. This photograph of the group was taken in October 1968 and includes some of the most prominent women in aviation. (Courtesy of Ann Wood)

each representing an area of major concern—airports and heliports, airmen, aircraft, flight standards and procedures, and education and public relations.

Based on the subcommittee reports, the group made many recommendations to the FAA. WACOA's suggestions, for the most part, were not new and this seemed to surprise outside observers. It was as if these individuals believed the feminine experience in aviation to be so different from that of men that women's ideas for change and reform would likewise be different. Concentrating on the basics, WACOA proposed parallel runways, suggested that there be a required clean-up of aircraft accident wreckage at airports, and argued for better cockpit design. The women knew that their proposals had been voiced before but believed they could take advantage of their unique position to stimulate the FAA into taking action on some of the most basic goals of aviation. The media was quick to promote the committee's more "feminine" proposal of airport beautification. The idea was in keeping with the Johnson Administration's (and specifically, in this instance, Lady Bird Johnson's) programs on highway clean-up, but it was not a major preoccupation of the committee.[34]

The degree of influence wielded by the committee varied. As an advisory group to the FAA administrator, the women could potentially affect policy. Even if the committee did not take any major actions, their very presence served as a constant reminder that women were involved in all capacities within aviation. The problem with the group was that it lacked the resources to do anything other than make suggestions. Though WACOA existed until President Carter dissolved it in January 1977, it became clear within a few years that WACOA's value was somewhat limited. It did create another means for top-level women to communicate with each other, and their collective voice did help encourage the FAA to make further advances in certain areas of general aviation. Its greatest weakness was that it tended to concern itself with changing programs without considering the dynamic social changes that were occurring. In other words, cleaning up runways was important, but it was not going to encourage more women to take flying lessons.

Major social upheavals occurred throughout the 1960s, and the decade saw a protracted reexamination of public ideals. The changing attitudes of Americans about work, pay, and what constitutes discrimination, began to be reflected in a demand for these changes to take place within the military. These debates helped open up the field of military aviation to women.

It was not an easy process. Enthusiasm for women and men in uniform was at an all-time low in the early sixties. Many studies from within the military sug-gested that no women should be involved in any facet of the armed forces. From the mid-1950s on, women had been gradually phased out of participation in many aviation functions, including serving as flight orderlies. Nevertheless, there were certain aviation specialties for which enlisted women in the Air Force were still preferred over men. These included work in air intelligence and defense combat control centers and in passenger air-transport operations. These were soon eliminated as the leadership decided that women's military occupations ought, in their words, be "in conformance with the 'present cultural pattern of utilizing women's services in this country.'"[35] This meant that in the Air Force enlisted women were no longer permitted to work in intelligence, information, or weather forecasting, or as flight attendants, or in equipment maintenance and control tower positions. Only 36 of 61 noncombat specialties remained open— 70 percent were in clerical work, 23 percent in medicine. In general, women were being slowly but systematically removed until, in 1965, only 30,600 women remained on active duty. Feminist groups took no action about this situation, partly due to the pacifist tone of their rhetoric and their general ambivalence toward the military.[36]

The key to the shift in attitudes was the escalation of the Vietnam War. The manpower problem was exacerbated by the increasing unwillingness of Americans to serve in a war they did not believe to be just. The draft was reinstituted in 1967. Concurrent with that announcement, plans were made to increase the number of women in the military.

The impetus for this action had come from the Defense Advisory Committee on Women in the Services in the spring of 1966. DACOWITS discovered that thousands of women had volunteered for service in Vietnam but had been turned away because of the recruitment ceilings. DACOWITS urged the military to define a more rational policy that would, at the very least, allow the recruiters to achieve the full two-percent quota of women that was permitted by law. DACOWITS did not stop there, however. With persistent and effective lobbying it helped to bring about the passage of Public Law 90-130, signed on 8 November 1967. The law accomplished two things. First, it removed all restrictions on the promotion of female officers. Previously women could achieve no rank higher than colonel in the Army, Air Force, and Marines, or captain in the Navy. Second, it removed all limitations on the number of female personnel who could be employed in the armed forces. Although the law was intended to bring greater equity to the career development of women in the military, it failed to resolve several issues. Two of particular concern to women in military aviation were the barring of women from the service academies and the statutory restrictions on women serving aboard aircraft engaged in combat missions and on ships of the Navy (except hospital ships and transports).[37]

The law did mean, however, that women were once again being utilized in "nontraditional" career specialties. In 1967, 1,223 WAVES, or about 20 percent of all women in the Navy, were assigned to naval aviation. At least seven of the 20 ratings open to women enlistees were directly connected with aviation. These included aerographers mates, radiomen, aviation electronics technicians, tradesmen, air controlmen, aviation storekeepers, and aviation maintenance administrators. Officers were also assigned to these fields. Ensign Sharon Fernando was the first WAVES officer assigned to work with a fighter squadron. She assumed the public affairs and educational services functions for VF-26, an instrument training squadron for Pacific Fleet fighter pilots. Training for this job included experience with the pressure chamber and ejection seat.[38] Another officer, Ensign Gale Ann Gordan of the Medical Services Corps, was the first woman in the history of the Naval Air Basic Training Command to fly solo in a Navy training plane. She received the flight training as part of the 111th Flight Surgeon Class in order to become an Aviation Experimental Psychologist. The only woman among 1,000 male students, she studied the same course they did, noting that her biggest problem was not the training but the flying clothing—all much too large.[39]

Marine Corps women were also involved in aviation medicine. This meant that women were routinely asked to complete the ejection-seat training program. During the 1960s women in naval aviation often underwent survival training, but in 1967 Airman Virginia Rookhaysen was the first woman to complete both the land and sea phases of the Survival School at Pensacola Naval Air Station.[40]

Minority participation was slowly increasing. In 1945 there had been two black women Navy officers; in 1968 there were 31, although this was still less than a half of one percent of the total number of female Navy officers. By contrast, the Air Force had approximately 218 black women officers during the years between 1966 and 1970, which amounted to about 4.6 percent of the total number of female AF officers.[41]

Between 500 and 600 Air Force women served in Southeast Asia during the Vietnam War. Operations officer Major Norma Archer gave the daily briefings on air strikes for the senior staff of the 7th Air Force. Air Force Intelligence was one of the most important functions performed by the women who served in Vietnam. In terms of numbers, the flight nurses were

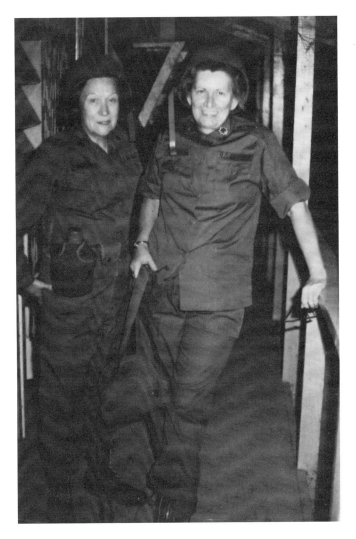

FIGURE 52.—Lt. Colonel Ann Johnson (left) and Lt. Colonel Yvonne Pateman, both USAF, at Tan Son Nuht, Vietnam, in October 1967. Both Johnson and Pateman were among the many women who served on active duty in Vietnam during the war. (Courtesy of Lt. Colonel Yvonne C. Pateman, USAF (Ret.), S.I. photo 85-15650)

the largest group of military women in aviation to serve in Vietnam. In fact, nurses represented the majority of American women in this location. One of the larger contingents was the 9th Aeromedical Evacuation Group, which arrived in Saigon in 1965. Many of the "ground" nurses were included in flying missions as part of air-evacuation teams to rescue casualties from active combat sites. Nobody seemed to object to the nurses' involvement in combat action. An important, though rarely acknowledged, part of the nurses' function was to boost the morale of the injured. Many women officers could not understand why it was acceptable for *nurses* to serve in the midst of the battle action, yet unacceptable for women in any other function.[42]

A turning point occurred when the Air Force began to make public its discussions on the potential use of women as pilots. Although the motivation was not completely apparent, part of the stimulus came from the public sector and part from internal pilot short-ages. In August 1967, *Air Force Times* published an article by Bruce Callander entitled, "Why Can't a Woman Be a Military Pilot?" In the article he wrote:

The Air Force has never seriously considered training women as military pilots. Its main reason is that all pilots are potentially eligible for combat assignments, and the Vietnam War has underscored this philosophy. AF has so far rejected even the idea of accepting female pilots for limited, non-combat duty although the approach has been offered as one possible solution to the pilot shortage.[43]

Callander made the case that historically women had proven capable as pilots. As we have seen, they had flown Air Force aircraft to set records and had participated in special tests. For example, Wally Funk, Jerrie Cobb, Jean Hixson, Ruth Nichols, Jacqueline Cochran, and many others all flew military jets during the 1950s and 1960s. Callander proposed that, while Congress might restrict combat flying, a compelling case was being made to use women for noncombat missions. Finally Callander suggested:

The ultimate question, however, may have little to do with any of the more practical factors of physical differences, mechanical aptitude, or emotional stability. For all their stress on equality of the sexes, Americans have perhaps the world's most over-developed sense of what kinds of work are proper for women.

The "front offices" of both military and commercial aircraft are likely to remain off limits to women for the foreseeable future. Still, male pilots should listen closely for an unfamiliar sound in the cockpit. This could be a feminine knuckle knocking ever so gently on the cabin door.[44]

The article was unusually progressive. During the 1960s, most women in aviation struggled to gain acceptance, let alone equality. Manifestations of the flying housewife still existed; programs were still sponsored for women because, as one *Flying* magazine article put it, "as mothers they would be able to influence their children."[45] However, greater numbers of women, both nonpilots as well as pilots, were beginning to recognize the role of aerospace technology and achievement in American society. Increased par-ticipation by women in aviation also resulted in empowerment—the possession of professional skills to accomplish a job and an awareness that these qualifi-cations entitled them to a place in the aviation world. Examples of women who were conscious of this experience during the 1960s were the members of the Women's Advisory Committee on Aviation and the Defense Advisory Committee on Women in the Serv-ices. The influence of these women extended farther as they came to represent the first frequent and persua-sive images of accomplished women in the field. Yet even these women were quick to recognize that this was just the beginning; they had yet to assume most of the real positions of influence on an equal basis with men.

During the 1960s, women in aviation were equipped with the legal tools of the Civil Rights Act of 1964, the Equal Pay Act, and Public Law 90-130, which removed career and manpower restrictions on women in the military. There was also a new-found sensibility with regard to the feminine potential. The concept of equal rights did strike a responsive chord in most Americans. Doors had been opened, and a new generation of pioneers was about to enter.

8. Women with "The Right Stuff:"
The 1970s

There is an implied promise, today, of equal opportunity for everyone in the Armed Forces. It is part of a new environment to recognize women as persons today, as well as pilots.

Antonia Handler Chayes

The title of Tom Wolfe's book, *The Right Stuff*, has become a popular description of the combination of ability, instinct, personality, and fitness that is supposed to characterize the supreme all-American male. Originally applied to the seven astronauts of the Mercury space program during the early 1960s, the expression also describes many of the women who were involved in aviation a decade later.

The 1970s encompassed a number of major international crises that caused domestic shock waves in American society. These ranged from the last years of the Vietnam War, to the Camp David Peace Accords, to the American hostage crisis in Iran. The Watergate scandal rocked the nation and challenged America's faith in government. It was a decade when there was a shift from the 1960s' style of effecting change through confrontation, to a new emphasis on seeking change through legislation and litigation.

Women in aviation in the 1970s, reflecting this pattern, often relied on the law and the American legal system to reinforce their claims to full and equal participation. In general, the presence of a woman in a nontraditional field no longer produced universal feelings of hostility. Americans, especially women, soon became enamored with the coverage of "female firsts." In aviation, that meant airline and military pilots, engineers, and corporate executives. The women were sometimes only token female employees, but never in any period since World War II had so many new fields opened up simultaneously, nor had so much attention been paid to women. In this respect women in aviation were very much like their counterparts of the 1920s and 1930s; their exceptional status attracted attention in itself and was also seen as a feminist statement. The women of the 1970s were the new pioneers. They were brave in the face of ceaseless media spotlights that sought to document the impact of female achievement (or perversely, failure) on America. Like the women flyers of the 1920s and 1930s, they were scrutinized and made involuntary role models for

a movement toward equal rights.

In 1972, approximately 156,000 women were employed in aerospace industries. They represented 17.2 percent of the total aerospace employment of 912,300. Gradually, both the number and percentage of women increased during the 1970s. In spite of industry stagnation and the slight decline of total aerospace employment figures, women made substantial gains. In 1979, 21.3 percent of the total industry employment (1,099,200) or 234,500 were women. This was a 50 percent jump in the total number of women in aviation in seven years.[1]

In manufacturing, women predominated in microelectronics assembly, which involved putting together the computer and electrical components that form the "nerve" system of modern aerospace equipment. This was the fastest growing area of aerospace manufacture. The first reason for the attraction of greater and greater numbers of female workers into aerospace electronics was that it was a light industry, and physical strength was not needed to participate.[2] Light industries such as electronics, food processing, textiles, and garment making traditionally have had higher concentrations of women employees than heavy industries.

The second reason for the increasing number of women in aerospace was related to the changes in college enrollment patterns. More women were attending college or other schools of higher education. Further, more women were majoring in the sciences and engineering. The number of women receiving bachelor's degrees in engineering tripled during the decade; even from 1967 to 1972 the number more than doubled, from 2,350 to 5,317.[3]

Women were being encouraged from all sides. Any industry dealing with the government was required to have an affirmative action plan to increase the presence of women and minorities and promote equal opportunity for all. Universities began actively recruiting female students. In 1973, the Massachusetts

93

Institute of Technology sponsored a symposium on "Women in Science and Technology," first to develop an understanding of why women had not enrolled in large numbers in the past, and second, based on that knowledge, to find ways of altering the pattern. The most important conclusion to come out of this conference, and many others like it, was the realization that in order to increase the number of women in science and engineering, proper primary and secondary school preparation in mathematics followed by similar training in college had to be provided.[4]

This message was echoed by women engineers. Kiki Fleck, a thermodynamics engineer at Lockheed who worked on the cooling of advanced avionics systems for the P-3 Orion and the S-3A Viking aircraft; Phyllis Veit, who worked at Aerojet Solid Propulsion in the Propellent Development Section; and Dr. Dora Doughtery Strother, Chief of the Human Factors Group at Bell Helicopter, all emphasized the importance of education. Each felt an obligation to take vociferous exception to the idea that women could not understand mathematics or become engineers; for example, Dr. Strother was a frequent speaker for Girl Scout troops, which she felt was a good way of communicating her thoughts to young girls.[5]

Olive Ann Beech still ranked as the pre-eminent woman executive in aviation management. In 1973 and again in 1978, *Fortune* magazine recognized her as one of the "Ten Highest Ranking Women in Big Business."[6] Beech was also a role model for other women in the aviation business like Athley Gamber, who ran one of the six largest small-aircraft service and sales companies in the southeastern United States. Gamber, who founded Red Aircraft in Fort Lauderdale, Florida (later part of Cigma Investments) with her husband in 1954, took over operations in 1968. When asked if she experienced any difficulties because of her sex, she replied,

The few women who are running businesses are highly capable. They've worked in the aviation game for a long time. And the biggest company in the United States, Beech Aircraft, is run by a woman. In a way, that sets the criterion for the whole industry.[7]

New laws and regulations that mandated affirmative action programs brought an influx of women employed by the federal government into aviation professions. In 1971 there were two important "firsts" for women working at the Federal Aviation Administration. On 4 April 1971, Ruth Dennis became the chief of the FAA's San Diego Flight Service Station, and on 16 May, Gene Sims became the tower chief of the Cuyahoga County Airport in Ohio. Both were the first women ever to hold these positions.[8] At the end of the decade the FAA rewrote its Aviation Career Services pamphlets to

FIGURE 53.—In 1974 Mary Barr became the first woman pilot with the Forest Service. (Courtesy of USDA, Forest Service)

include only gender-neutral terms; for example, ramp serviceman became serviceperson, and "he" and "she" were used jointly throughout the text. In addition, the FAA also reprinted for public distribution a series of magazine articles about women. The four reprints included information on women in Navy and Air Force aviation, as well as more general surveys. Yet, for all the official "equality," numbers and attitudes changed slowly. During the late 1970s women accounted for only about five percent of the total number of air traffic controllers. Lynne DeGillio, who was an air traffic controller at John F. Kennedy International Airport in New York, described the tower as retaining its "locker room ambience" despite the addition of women.

Throwing insults or teasing, sometimes on the most juvenile level, is a release for them—a time when they're not thinking about airplanes—and they do it at every chance they get. Everyone's personality is magnified in that tower, and you can't let it get to you. You have to become one of the guys to get along in what appears to be their world. At the same time, you have to hang on to your femininity.[9]

DeGillio resented the outmoded attitude of her colleagues, and felt tension between being part of the group and just being herself. That additional stress was a common experience for women in aviation, although it was not limited to the airport tower. For example, other FAA women who worked as accident inspectors, flight examiners, or engineers all reported similar feelings. However, the government was better than much of private industry, in that at least there had been a sustained, if small, presence of women in most fields of federal employment for some time.[10]

United States airlines, on the other hand, had experienced a long hiatus between the first woman commercial airline pilot (Helen Richey in the mid-1930s) and the next in the 1970s. According to unconfirmed Soviet reports, in the 1950s Aeroflot hired its first female commercial airline pilot. In 1961, Scandinavian Airlines System hired Turi Wideroe, and she became the first woman pilot for a major airline outside the Soviet Union. In the early 1970s, women pilots were flying for many foreign airlines, including Sabena, Mexicana, El Al, and Air Inter of France. It was not until 1973 that the first American woman was hired to fly for a major airline. In January 1973, Emily Warner became first officer on a Convair 580 for Frontier Airlines, opening up for American women one of the last sex-segregated aviation occupations in the nonmilitary aviation industry.[11]

Warner was 33 when she was finally hired by Frontier in 1973. Flying since the age of 17, she had held jobs as diverse as flying traffic reporter, FAA examiner, and flight school manager. She was the chief pilot for the Clinton Aviation Company in Denver when she first applied to Frontier in 1968. The employment opportunities for pilots were limited, but when Frontier decided to recruit a new group in 1972, Warner was there with application in hand. She was very persistent, and when she was finally hired, a spokesperson for the airlines said: "We couldn't think of any reason not to."[12] In 1978, there were about 50 women out of a total of 38,000 pilots flying for United States airlines. Except on very small feeder lines, most were flight engineers or second officers, if for no other reason than they had not accumulated sufficient hours and seniority to become captains.[13]

The women wanted to be pilots for the same reasons as men. First, they loved flying, but they also loved using their talents in respected, well-compensated work and they loved the feeling of belonging that the airlines try to create. Many also had strong family connections with flying. Often they had relatives who flew or worked for the airlines.

The obvious question that was being asked in the late 1970s was "why weren't women hired before 1973?" The airlines claimed that qualified women had not expressed interest or applied. Economic and marketing priorities were more decisive factors, however. Highly regulated by the federal government and unable to manipulate either routes or fares, the airlines competed for passengers through service. Image was extremely important. In the early 1970s, there were still significant numbers of Americans who had never traveled, or traveled only once, on an airliner. Even for trips over 200 miles, most Americans still drove their cars. Consequently, the airlines were still very much wedded to a public relations program of safety and convenience. They deliberately cultivated the image of the pilot as "father" and were uncertain as to whether or not putting women in the cockpit might instill new fears in passengers' minds.[14]

The airlines had little experience with women as career professionals and did not see them as primary wage earners: men needed to support families, women did not. The airlines argued that women would soon quit to start a family, thus supposedly validating the notion that hiring women would not be cost effective. The jokes about "women drivers" were rife in the back rooms of executive suites and behind the closed cockpit door.[15]

The airlines were dependent on the Civil Aeronautics Board for subsidies and protection from competition. In the 1970s, the airlines had to become "Equal Opportunity Employers" if they were to receive federal money. Further, the equal rights movement constantly lobbied the companies. The movement had much to gain if women were hired. Airline pilots were highly visible and well regarded. The National Organization for Women wanted visible role models of women in nontraditional careers to demonstrate the legitimacy of its philosophy and goals.

Regulations and enthusiasm aside, the airlines did not have hordes of women pounding on their doors. The requirements for application were stiff. An applicant had to be at least 25 years old; have four years of college; a commercial license; an instrument rating; an airline transport rating; 1,500 hours flying time, including both night and cross-country flying experience; and a first-class medical certificate. In addition, applicants had to be familiar with navigation, FAA regulations, and the principles of safe flight. They were required to have exceptional hearing and sight and meet physical requirements such as minimum height (five feet, six inches for American; five feet, eight inches for Delta). Fifteen-hundred hours was a considerable amount of flight time. Outside of the armed services, preparation for a career as a pilot meant a substantial financial investment. Women had just begun to fly in the military; a number of years would have to pass before this could be used as an avenue of career preparation.[16] In addition, it should be noted that these requirements resulted in an even smaller presence of minority women working to attain employment with the commercial airlines. There were only 110 black pilots working for the airlines in 1978, when Jill Brown became the first black woman to fly for a major United States commercial airline. Brown, who worked for Texas International Air, had been the first black woman to enter the Navy's flight school (in 1974), though she left the military after six months of training.[17]

The airlines did not have a consistent policy to encourage the presence of women, nor did they have one that might alter the ratio of white women to women of color. The female pilots who advanced in their careers did so in spite of the existing conditions. Likewise, women who sought to work in management and other executive positions at the airlines experienced a similar inconsistency in the airlines' reactions to their career potential. Nevertheless, there were some highly successful female airline executives.

United Airlines had several women serving in high-level management positions. Julia McMurray was named manager of the women's market in 1972. Because the airlines were so conscious of their market image, McMurray's job was to plan and direct programs geared for women travelers. Barbara Allen headed the section that included computer and data-processing systems. Marge Segal worked as an account executive, handling United's convention sales. Peggy Ann Moore was an assistant flight dispatcher based in San Francisco. Most women at United who worked in management positions during the 1970s were in the public relations department. No women had become chief executives, but they were slowly beginning to advance in the corporate hierarchy.

A good illustration of this evolution can be seen in the career of Ann Wood. She was one of the original 24 women picked by Jacqueline Cochran to fly with the Air Transport Auxiliary in the United Kingdom. During the postwar period, she worked for America's first civil air attaché of the United States embassy in London. During the 1950s she was director of public relations for Northeast Airlines in Boston. In the early 1960s, Wood served as the project officer in a Massachusetts Mass Transportation Demonstration Project. Between 1966 and 1977, she worked at Pan American World Airways in public relations. She left the company as staff vice president to begin a new position as assistant

FIGURE 54.—Ann Wood receives a citation from the Federal Aviation Administration for her work on the Women's Advisory Committee on Aviation. Wood was a ferry pilot for the British Air Transport Auxiliary during World War II and later worked in public affairs for several airlines, including a term as staff vice president for Pan American from 1966 to 1977. (Courtesy of Ann Wood)

to the president of Air New England, where she continued her work as liaison between the New England community and the federal government.[18]

The other principal group of women connected with the airlines was still the flight attendants. The single most important development in the 1970s was a shift in corporate and public opinion, so that the attendant was now seen as a career professional, not simply part of the service personnel. This change paralleled the increase in the average tenure of a flight attendant from two years to seven. There were indications, such as a high level of job satisfaction, that this average would continue to rise during the 1980s. Even with this change, flight attendants were still perceived as "attractive, mature personalities with glamorous jobs." True as this perception was, the image had changed over the years. Class and age requirements became less rigid. Applicants were still expected to be attractive and congenial, but it was a "clean and wholesome rather than glamorous appearance [that was] desired."[19] By 1979, starting salaries ranged from $8,000 to $17,000 annually, plus a variety of benefits such as reduced travel costs, insurance, annual leave, and retirement plans.

Salary, benefits, and working conditions were negotiated by union representatives at contract meetings. The three main unions were the Association of Flight Attendants (AFA), the Teamsters, and the Air Transport Division of the Transport Workers Union of America (TWU). The largest of the company unions was the Independent Union of Flight Attendants (IUFA), which represented attendants with Pan American. Total combined membership of the independent unions was 19,000. In contrast the AFA, the largest of the multi-carrier unions, had about 22,000 members. The Teamsters had roughly 5,100 flight attendant members and the TWU about 7,100.

The AFA was the new autonomous union of flight attendants that had grown out of ALPA's Stewards and Stewardesses Division (S & S Division). It was the largest union and represented approximately half of the unionized flight attendants in the United States. Under the original S & S Division arrangement, it had been possible for a flight attendant to be elected as president of ALPA. When the ranks of attendants began to swell in the early seventies, it was anticipated that the S & S Division's membership would surpass that of the pilots. Fearful that the women (S & S was predominantly female) might elect one of their members as president and that they would lose control of "their" union, many pilots, in conjunction with the ALPA leadership, began to encourage the formation of an independent union. The attendants had mixed views on the situation. Some wanted total independence, others wanted to preserve the status quo. In October 1973, J.J. O'Donnell, president of ALPA (head of the pilots), and Kelly Rueck, vice president of ALPA (head of the S & S Division), met to discuss the creation of a new organization. The proposal created separate but affiliated pilot and flight attendant organizations, which would continue to share administrative offices, supplies, and legal counsel.[20]

The agreement was ratified in November 1973, and preparations were made to seek the National Mediation Board's recognition of the newly created Association of Flight Attendants (AFA). The board ruled in December 1974 that a "simple majority of one vote more than 50 percent," by the attendants of each separate airline that had formerly been a part of the S & S Division would serve to legally transfer bargaining representation from ALPA to AFA. Over the next five years attendants at each of the 16 carriers voted their approval to give AFA bargaining rights in contract proceedings.[21]

FIGURE 55.—Kelly Rueck was elected vice president of the Air Line Pilots Association for the Stewards and Stewardesses Division in 1970. She later became one of the chief architects of the new Association of Flight Attendants, serving as AFA's first president until January 1976. (Courtesy of the Association of Flight Attendants, S.I. photo 86-11867)

Contract negotiation took on a new air as the women reorganized their methods and became as professional as possible. They took classes on negotiation techniques and took the time to research the history of past negotiation efforts. This involved interviewing all of the participants and re-reading all of the previous contracts and their supporting documentation. For the first time the flight attendants truly had an understanding of the attendant's place within the corporate structure. The first test came with United Airlines. Initially the company had agreed to the requests of the negotiation team, but problems quickly developed. Essentially, the flight attendant team wanted to be

I'M GONE 273 HOUR PER MONTH
I'M PAID FOR 85 HOUR

Piedmont

PIEDMONT
Flight Attendants
Serving You Without a
CONTRACT
since May 1, 1975

FIGURE 56.—Piedmont flight attendants participate in a strike caused by stalled contract negotiations. Strikes were part of the new negotiating tactics used by the unions to force a company to make concessions. (Courtesy of the Association of Flight Attendants, S.I. photo 86-11858)

leaders and the United bargaining team improved communications in order to make sure all understood the goals as well as the means that would be necessary to accomplish those ends. The women learned not to place artificial restrictions on themselves based on societal expectations and unquestioned stereotypes.

The airlines experienced considerable growing pains as they began to relinquish the old notions of what the job of a flight attendant really represented. It was no longer primarily an adventure, but was essentially work. As personnel turnover rates dropped during the 1970s, the airlines slowly recognized that they would have to regard the attendants as more than short-term employees, and accord them a new and greater status within their corporate hierarchies.

recognized and treated professionally (as were ALPA and the other unions). United was completely surprised by the flight attendants' new tactics, rather than by the actual contract demands. The women persisted, but instead of gaining acceptance they were branded as radicals. Company officers claimed that members of the team were out of touch with the United flight attendants, that the negotiaters were not like the other "normal" attendants. This was an attempt to explain the company's failure to completely control the proceedings.[22]

The women ultimately succeeded both in winning contractual gains and in unifying the members by a process of intensive education. The flight attendant

afa
FOR
ERA
association of
flight attendants
AFL-CIO

ASSOCIATION OF FLIGHT ATTENDANTS

FIGURE 57.—During the 1970s and 1980s, the Association of Flight Attendants actively worked in support of the proposed Equal Rights Amendment. (Courtesy of the Association of Flight Attendants, S.I. photo 86-11862)

The military, in contrast, was on the verge of a public relations campaign to convince young women that they should join up precisely because participation would mean "adventure" and "not just a job." The creation of an all-volunteer force in 1973 filled recruitment officers with anxiety. They were uncertain about their ability to meet the strength requirements being set forth by the Department of Defense. Concurrent with this dilemma was the congressional debate on the Equal Rights Amendment. One of the central issues in those discussions was the role of women in the armed forces. This triggered a renewed awareness among the manpower strategy experts at the Defense Department as to the potential value of women. In any event, there seemed to be a good possibility that the military would soon be required by law to use women without restriction. Numerous special studies on the deployment of women cropped up, each coming to the conclusion that there ought to be an increase in the number of female personnel. George A. Daoust Jr., Deputy Assistant Secretary of Defense (Manpower Research and Utilization), asked each service to develop contingency plans for the increase in the number and use of women in anticipation of a law such as the Equal Rights Amendment being passed. Daoust also indicated that each branch, except the Marine Corps, should plan to double its female ranks by 1977. The Marines were only asked to produce a 40 percent increase.[23]

Once the individual services began to examine the issue seriously, their contingency preparations became effective plans of action. Throughout 1976 there was a 20 percent growth rate for women in the military. After that year the rate slowed to five percent, as the armed services sought to determine the impact this enlarged percentage of women would have on operations and effectiveness.

The women who had signed on found there were a multitude of new opportunities in aviation: Reserve Officer Training Corps programs were opened to women, restrictions because of sex were removed from many military occupations, the academies became coeducational, the first women were promoted to the rank of general or admiral, and most significantly, the first women were being trained as pilots. What was so important about this last change was that flying was, and continues to be, one of the essential activities of military air service. Even if the numbers were small, the advent of women's participation was symbolic of a movement toward full integration.[24]

In January 1973, Secretary of the Navy John Warner announced that eight women had been selected to enter the Navy flight-training program at Pensacola (officer training for women would be held at Newport,

FIGURE 58.—Corporal Smith, USMC, working as an avionic electronics technician at Marine Corps Air Station, Cherry Point, North Carolina, in 1979. (Courtesy of U.S. Marine Corps, DoD Still Media Depository, DM-SN-82-07766)

Rhode Island). Their course would be identical to the men's, because the goal was to determine the "feasibility of using women in non-combat flying billets in helicopter and transport squadrons."[25] Eighteen months later, in 1974, six of the eight had earned their official Navy Wings of Gold: Jane M. Skiles O'Dea, Barbara Allen Rainey, Judith A. Neuffer, Ana Marie Fuqua, Joellen Drag, and Rosemary Conaster. Rainey, who was the first woman to receive Navy wings, commented on the experience: "Everybody goes through a stage of being depressed. The hours are long and the work is hard. You sometimes think, 'What's the use? There's so much to learn. I can never do it.' I think all students go through that."[26] However, the women faced the additional pressure of questions, spoken and unspoken: "Can they adjust? Can they do it?" The first class demonstrated that women could learn to be good military pilots. The personality traits and skills needed to complete Navy flight training were shared by women and men alike. Based on that knowledge and the good

FIGURE 59.—Flight Nurse Captain Cathy Young, USAF, was a member of the 375th Aeromedical Airlift Wing. Here she directs an aeromedical unit to move emergency vehicles from the path of the C-9 Nightingale aircraft, which is about to take off. (Courtesy of the U.S. Air Force, DoD Still Media Depository, DF-ST-82-04164)

experience with the first group of women, a second class was authorized in 1975.

The authorization did not come without controversy. Many in the Navy felt there simply was not a place for women pilots. Their objections were voiced in influential quarters and in sophisticated rhetoric. The excerpt below is taken from an essay in the *U.S. Naval Institute Proceedings:*

Until such a time as women are legally and physically able to occupy any job in any location under the same conditions as men, a most unfair and discriminatory practice exists and the Navy must stop it. Let's chalk up the initial input of female aviators to an interesting experiment in equal opportunity that didn't really provide equal opportunity and stop at that.[27]

Women in aviation have usually found that participation has hinged on proof of ability. The military leadership was convinced the passage of the Equal Rights Amendment was imminent; hence it wanted to determine the potential of women in various roles previously closed to them. Many still defined women in terms of their limitations, however. This created an environment of tension for women, although with each successive (and successful) class of women, a portion of the conflict abated.

The third class followed quickly on the heels of the second. It was also the first to have the Aviation Officer Candidate School phase conducted at Pensacola. (The previous two had held this phase in Newport before transferring the women to the Florida air station for flight training.)[28]

In 1973, shortly before the flight programs for Navy women began, 12 percent, or 713, of the enlisted women in the Navy worked in aviation specialties. That same year the WAVES corps was abolished, and women were brought into the regular Navy. Instead of being WAVES or WAVES officers they were simply officers and enlisted women in the Navy. By mid-1977 there were 4,000 enlisted women involved in Naval aviation in 24 separate ratings. In addition more than 100 officers and 1,000 enlisted women worked in aviation as part of the Naval Reserves. This growth, matched by the successful adaptation of women, has obviously meant many changes for the Navy.[29]

Yet the key question regarding women in combat remained. When asked why she wanted to participate in combat, Lieutenant Junior Grade Rosemary Conaster replied:

Why do I want to go to a tactical squadron, to fly off a boat, and perhaps be shot at?

My reasons are the same as those that have always attracted men to Naval Air. It is because I have experienced the satisfaction of the first step—winning Gold Wings—and I want to continue to succeed at what is the most demanding form of aviation. I want to become a full professional in my chosen vocation.[30]

To be a fully accepted member of the Navy, women recognized they would need to share in the risks and responsibilities.

The Army began to include women in flight-training programs at almost the same time as the Navy. Lt. Sally Murphy received her wings in 1974, becoming the first woman aviator in the Army's history. Unlike the Navy, which had a separate class of women, Murphy was simply one of 25 members of Officer Rotary-Wing Class 7414 at the Army Aviation Center based at Fort Rucker, Alabama. Like many other women pioneers in military aviation, Murphy stressed that, in spite of being the first and only woman, her colleagues were very accepting of her. "I was allowed to maintain my femininity, but the men did not pamper me nor give me special treatment—they maintained a perfect balance in our relationship."[31] Women were trained in many of the Army aviation specialties, including maintenance. They attended "Jump School," learning the skills of an Army parachutist. Several women encountered difficulties in learning certain specialities because of their lack of experience with tools and machinery prior to attending Army technical schools. The Army had not anticipated this situation and did not make any curriculum adjustments to compensate, despite explicit congressional instruction to do so. The attitude was that women would have to adjust to the existing standards the same way men did, whatever the consequences. This did affect the participation of women, as some dropped out of the specialty, and others were discouraged from even trying.

This was true in the other services as well. Navy airplane captain Ellen Mahoney stated:

On my first plane I was plenty nervous. For the first couple of weeks I became a hermit. I was so tired when I got home, I just died. I'd ask the guys if they felt the same way and they would say "no." They always seemed to feel great. Now I've found out that it takes a lot of getting used to.[32]

Mahoney really liked her job. She was quick to praise her colleagues and officers, although she still had to prove herself to the men. Her boss commented, "When I heard I was getting a girl on my line, all I could think was 'Why me?' But she sure has turned my attitude around."[33] The combined pressures of learning a new job as well as having to prove oneself were overwhelming for some women.

The Coast Guard, which during peacetime is a civilian agency under the control of the Department of Transportation, also trained its first woman helicopter pilot in the seventies. Janna Lambine applied for flight training while attending the Coast Guard's officer candidate school in Yorktown, Virginia. On 4 March 1977, she earned her wings and was designated a Naval Aviator at Whiting Field Naval Air Station. Lambine was definitely a token woman, as the Coast Guard had shown only a limited commitment to training women aviators. The organization was sensitive to the public pressures that demanded at least the appearance of integration. Lambine experienced an isolation that was typical for the first representative of a particular minority when entering the profession of military aviation.[34]

The Air Force took longer than either the Navy or the Army to incorporate women into its flight-training programs. It maintained that even though women had completed their programs in the other services, the reason for exclusion was not that the Air Force did not believe women could be pilots, but rather that the legal restrictions on women in combat made it difficult for them to justify an extensive program to bring in women. The Air Force claimed that all pilots were combat pilots, though many pilots might "spend their entire careers as instructors, transport pilots and other assignments without drawing a combat tour."[35]

Eventually the Air Force began to include women in other aviation programs that included flying experience. For example, in July 1975, Captain Jane Holley graduated as the first women flight test engineer.

Coincidentally her class was also the first held at Edwards Air Force Base. She went to school side by side with the Air Force test pilots. In the 44 week course she had over 100 hours of flight time. Her one comment upon graduation was: "I only hope I have done well enough so that other women will be welcome at the school."[36]

In November 1975, there was a major change in Air Force policy on this question. During his farewell address at Andrews Air Force Base, Secretary of the Air Force John McLucas, to the surprise of many in the audience, announced that women would soon begin training as pilots. McLucas noted that the Air Force was considering the development of a two-track pilot program, combat and noncombat, but was not convinced this was the only way to train women.[37]

The first official reports on the plan came on 10 December 1975. *Air Force Times* noted that representatives from the Air Training Command and Military Personnel Center in Texas had met with Air Force legal, logistics, personnel, and medical staff to discuss anticipated problems and coordinate the entire operation. One result of the meeting was the decision to permit women who would be entering the Air Force Academy the following fall to take the T-41 training program, if the general program was successful.[38]

In January 1976, the Air Force announced its program goals. These were (1) to determine how women pilots could be used within the Air Force outside of combat missions, and (2) what curriculum modifications would be needed if a permanent course for women were added. The Air Force wanted to determine potential career paths for women that included a list of positions for which they would be eligible.[39] At the same time, the Air Force called for applications. By April the committee had decided which airplanes the women could fly and where the training would be held—Williams Air Force Base, Arizona. The Air Force also decided to eliminate the helicopter components of the experiment, because they felt the Navy and Army had fully demonstrated that it was possible to train women to fly rotary-wing craft without any problems.

Following the initial decisions on the course structure, the Air Force turned to the logistical details of the plan. The first major problems they confronted were related to physiological and medical issues. Probably the second greatest barrier to women pilots in the military was the physical dimensions of the cockpit and the complete failure of aircraft designers to apply knowledge about female physiological characteristics, such as average height, sitting position, reach, and eye height. The design of a cockpit could make the difference between safe and unsafe operation of an aircraft. The average stature of pilots, a determining factor in safe cockpit design, would be reduced if the female physique were taken into account. Ignoring this change would introduce an element of danger with regard to existing aircraft. It was recognized that the permanent inclusion of women would necessitate changes in everything from cockpit configuration to uniform design and supply. That recognition triggered the surfacing of latent resentment toward women for upsetting the system. Finally, there was the debate over whether women pilots would be allowed to take birth control pills (up to this point pilots could not be on medication of any kind).[40]

In August 1976, members of the first two women's classes were named. During the press announcement ce official commented that "although the AF refers to the female pilot training program as a 'test program,' some AF officials see no reason why they should not be successful. Women have been flying for almost as long as men."[41]

Before the first half of the first group had even finished training, the Air Force decided to continue the program. Another nine women would begin training in February 1978. The reason for this was the positive reactions of the instructors at Williams. Lieutenant Colonel C.T. Davis, operations officer for the 96th Flying Training Squadron reported that

a lot of people had a lot of ideas that women wouldn't be able to hack it because of their lack of physical strength, because of inadaptability to stress, because of this, because of that. So far our experience has been that it hasn't changed things at all. The women are going through exactly the same training as the men and are hacking it just as well.[42]

When the women were asked in public about how well the men accepted them, they always replied positively. Privately, many expressed a fear that any criticism on their part might end the program.[43] There were difficulties, often created by the extra attention given to the women, yet it was that attention which helped women bear the additional pressures of success. Captain Kathy La Sauce commented: "When you step out of the airplane and take off your helmet, the transient maintenance guy almost falls over backwards. That sort of keeps us going."[44] There was a great deal of pressure to be good, to succeed lest the door be shut for other women.

Women pilots faced a very uncertain future. They were doing what they loved best, yet their career options were rather limited. The Air Force was subtly letting women know that it was not "women's lib," but rather a shortage of qualified men that had led to the decision to train them. The implication was that if more men had been available, the women would not have been pilots. In light of their continued progress,

however, the Air Force began to want to remove some of the combat restrictions placed on women so that more of them could be used. It strongly supported requests to Congress in 1977 and 1978 from the Department of Defense that would have allowed the secretaries of each service to establish duty assignment policies. The proposed amendment to Title 10, Sections 6015 and 8549, would have specifically allowed the assignment of women to noncombat ships and aircraft. The wording of the amendment was such, however, that the possibility for ending any and all restrictions on women in combat would also have existed.[45] This ammendment was not successful and the original policy remained unchanged in 1985.

The other major flying career newly opened to women in the Air Force was navigation. The first class of six women entered school at Mather Air Force Base in March 1976. The course lasted 33 weeks and the women who graduated were to have the opportunity of flying in KC-135, WC-130, C-141, and WC-135 aircraft. Captain Bettye Payne, who graduated with this first class, said she applied after receiving encouragement from OV-10 aircraft pilots in Korea. She wrote a brief article in which she said, "The best thing about being a navigator [was] the challenge of flight. No two flights [were] alike and I liked that element of surprise."[46]

There were two other major congressional battles dealing with women in military aviation. The first was the decision to open the service academies to women. Unlike the Navy and the Army, the Air Force did not express unanimous opposition to the idea of admitting women. WAF Director Jeanne Holm went on the record as supporting the idea (her counterparts in the other services did not agree), and Secretary of the Air Force Robert Seamans also indicated his approval when he said the Air Force would accept women nominated by Congress if funds for new facilities (housing, bathrooms, locker rooms, etc.) were provided. In general, however, the Department of Defense was opposed to the idea and made every effort to lobby against the bill. Jacqueline Cochran took this occasion to testify forcefully against admitting women to the academies, because she felt the proper and primary role for women in American society was mother and housewife.[47] Despite her own experiences and those of the Women Airforce Service Pilots, she felt there was no role for women in the military during peacetime. It is alleged by several of Cochran's colleagues that she was prevailed upon by individual members of the Air Force to take the stand she did. Congress refused to listen to such arguments, and in the spring of 1975, the mandate was issued. Public Law 94-106 decreed that as of September 1976, the academies would admit women. The Air

Force took the most active role in encouraging women to apply, and after the first summer of training, the Air Force had the lowest attrition rate for women of any academy.[48]

The importance of the decision to open up the academies is that it provided another avenue for women to obtain leadership positions. The academies' goal had been to prepare men to be combat officers. They also provided an elite corps, many graduates of which would ultimately attain the highest ranks in their respective service. The inclusion of women naturally altered the educational process, but it also opened up another set of options for a young woman who wanted to be a career military officer. The women graduates would be well placed to influence the participation of women in military aviation. Their mere presence had a liberalizing and liberating effect on their male colleagues. Further, academy graduates (especially in the Air Force) who became pilots stood the greatest chance of one day being promoted to general.

The second important congressional battle of the seventies regarding women pilots dealt with individuals at the other end of the chronological spectrum. This was the newly reopened lobby for the provision of official recognition and veterans' benefits for the WASPs. The first attempt was a bill introduced by Representative Patsy Mink in May 1972. It failed even to be considered by the House Veterans Affairs Committee, largely due to the opposition of the Veterans Administration (VA). On 26 March 1975, Senator Barry Goldwater introduced another bill on behalf of the WASPs. Like Mink's bill, Goldwater's also died in committee, again blocked by the VA. In May 1975, Representative Mink made yet another attempt. The bill was killed in committee a third time. Then in September 1976, there was a bill before Congress for veteran status for Polish and Czechoslovakian citizens who fought with the Allies and who later became United States citizens. Senator Goldwater added an amendment to the bill that provided like recognition for the WASPs. The Senate passed the bill with this amendment, but the House steadfastly refused. With the removal of the amendment the WASPs lost yet again.[49]

The following year, 1977, was declared the "Year of the WASP" in hopes of winning recognition through an intensive program of lobbying. Representative Lindy Boggs of Louisiana and Senator Barry Goldwater sponsored bills in the House and the Senate respectively. It was the only bill ever to be cosponsored by every woman member of Congress. The WASPs received the support of the military news weekly *Stars and Stripes,* along with that of Antonia Handler

Chayes, Assistant Secretary of the Air Force for Manpower, Reserve Affairs and Installations. The Carter Administration, however, following the recommendation of the Veterans Administration, did not support the bill at all. The Veterans Administration argued that if status were granted to the WASPs, then precedent would be set for innumerable civilian wartime organizations to make the same request. Like its predecessors, the 1977 House bill would have died in committee but for the inclusion of one small item in a packet of documents forwarded to Representative Olin Teague. Teague staunchly opposed the bill, but when he saw that the copy of Helen Porter's WASP discharge was identical to the official World War II Army honorable discharge certificate, he changed his mind. Teague acknowledged, for the first time, that the women had been de facto military personnel. With his support the bill passed easily on November 3. The next day the Senate also passed the resolution. On 23 November 1977, President Jimmy Carter signed the bill into law.[50]

The support of Secretary Chayes had been critical to the successful passage of the law. Chayes believed that the WASPs were an important part of Air Force history. Their recognition, she believed, would be significant in the ultimate integration of women pilots in the Air Force. Defeat of the bill would have sent the wrong message to women who had just started pilot training in the Air Force and would have fueled opposition by military leaders who did not want women fliers.[51]

The media concentration on so many women pilots from a previous generation did inspire many young women military pilots. Through newspapers and television programs, the WASP issue brought attention to all women in aviation. In 1970 there were 29,832 licensed women pilots. Nine years later there were 52,392, an increase of nearly 80 percent, in contrast to an 11 percent increase in the number of male pilots.[52]

The vast majority of the 52,000 were general aviation pilots. In the early seventies, two scientists began to conduct studies to determine the personality characteristics of general aviation pilots. It was the first such research effort to include women. The first hypothesis the scientists formed was that female pilots possessed a personality distinct from the standard United States female adult population. The second was that there was a "pilot's personality" that transcended sex differences. The research strongly supported both hypotheses. The personality of female pilots was called active-feminine (in men active-masculine) and was described as a woman who was "courageous and adventuresome, one who is oriented toward demonstrating competency, skill, and achievement; one who finds pleasure in mastering complex tasks; and one whose sexual orientation is decidedly heterosexual."[53]

The purpose of the study was to collect data on women in aviation as part of a program of increased aviation safety through human factors research. It was demonstrated that there was no scientific evidence of "dramatic" differences between men and women pilots. In fact, according to the study, they were very similar in character. Critical factors that the test failed to explore were whether race, ethnicity, or economic status made any difference in pilot personality traits. It was impossible to tell whether or not the subject pool reflected the general United States population. Because of the considerable expense involved in flying, the economic status of the subjects would have been illuminating. Had public policy or engineering decisions been made on the basis of this data exclusively, the net effect might have been to perpetuate the status quo, which would have meant the continuation of general aviation as the sport of wealthy white men.

Returning to the original theory about a "pilot personality," none fit the profile better than Mary Gaffaney, who was the subject of a feature article in *Sports Illustrated* in March 1973. A native of Miami, Gaffaney had been a professional pilot and aerobatic champion for 26 years. She was an instructor for helicopter and sailplanes as well as for airplanes. Five-time U.S. National Women's Aerobatic Champion, she also won the Women's World Aerobatic Championship in 1972 at Salon-de-Provence, France. Fellow contestants uniformly praised her abilities. "The consensus among Mary's past and present peers is that she has succeeded in aerobatics because she works hard at it, and since she is in the flying business, she can afford the game."[54]

Aerobatic flying skill took countless hours to perfect. It required constant access to an aircraft and the financial resources to pay for the costs connected with flying. Gaffaney was the co-owner of a flying school. She was also a skywriter (the first woman to enter this elite form of precision piloting), which gave her additional practice time.

Gaffaney's flying skill was not limited to fixed-wing

FIGURE 60.—Mary Gaffaney won the U.S. National Women's Aerobatic Championship five times. In July 1972 at the Salon de Provence, France, she became the first American woman to win the Women's World Championships. (S.I. photo 86-12154)

craft. She was also a helicopter instructor, a Whirly-Girl, and the third pilot for the United States team at the Second World Helicopter Championships held in July 1973 at Middle Wallop, England.

The Whirly-Girls were well represented at the World Championships, which were open to teams of men and women (both mixed and single sex) from all nations. Whirly-Girls from several countries served as individual pilots. Amidst all of the Army, Air Force, and Navy teams from around the world, only the United States fielded an all-woman team, made up of Ilovene Potter, Betty Pfister, Betty Miller, Charlotte Graham (now Kelley), Mary Gaffaney, and Jean Tinsley. The team delegate and judge for the United States was Jean Ross Howard.[55] All of the women were exceptional pilots, and although the team did not win any of the competitions, their participation was a strong tribute to the Whirly-Girls, who had provided the leadership needed to organize a team and had secured full sponsorship from Bell Helicopter. The team illustrated the new level of acceptance and respect being accorded to women in aviation—both nationally and internationally.

The Ninety-Nines likewise achieved wider recognition during the decade. The scope of their activities and membership broadened considerably. During the 1970s, the Ninety-Nines became very involved in the National Intercollegiate Flying Association, contributing money and time (for example, judging competitions) to encourage young women (and men) in aviation. Further, there was a strong emphasis on humanitarian work. The Ninety-Nines worked with groups like Happy Flyers, Flying Samaritans, Blood Flights, and the Civil Air Patrol. A prime example of this orientation is to be found in Nine-Nines member Jerrie Cobb. She had been a NASA consultant and potential astronaut candidate in the 1960s. Afterward she devoted her life to working as a jungle pilot in the Amazon River area of South America, committing all her personal resources to flying doctors, missionaries, anthropologists, and supplies in to the Indian tribes, as well as providing an air ambulance service for Indians needing emergency medical attention outside their home area.[56]

The changing composition of the membership transformed the Ninety-Nines. The advent of female military pilots, commercial airline pilots, and an increase in the number of women professionally involved with aviation gave the organization a new vitality. Growth and development were inevitable.

The last All-Woman Transcontinental Air Race was held in 1977. The demise of the race was caused by rising costs, diminished corporate sponsorship, and new levels of air traffic congestion.[57] Competition in the

FIGURE 61.—Jean Ross Howard, Director of Helicopter Activities at Aerospace Industries Association of America, Inc., is one of the foremost women in aerospace in the 1980s. She is also the Executive Director of the Whirly-Girls, an international organization of women helicopter pilots, which she founded in 1955. (Courtesy of Jean Ross Howard, S.I. photo 86-12198)

air was still important, as the continued existence of other races, including the Angel Derby, proved. Women's air races were described as "joyous aviation community effort.... Everybody in this adventure must win, whatever winning is, and the afterglow spreads through the entire aviation community."[58]

Other activities absorbed the energies of the organization in place of the AWTAR. In particular, the Ninety-Nines became committed to a program of aviation education. The international officers of the organization wanted to make people everywhere, especially the young, interested in aviation. They appealed to a broad base, because it was important to the Ninety-Nines to demonstrate that every woman was welcome to participate. Realizing the value of example, they never missed a chance to publicize the outstanding accomplishments of their members.

During the 1970s, the Ninety-Nines were transformed into a network of both professional and nonprofessional women in aviation.[59] They provided support and enthusiasm for each other, but their greatest contribution was to encourage women to move beyond the glow of being first to do this or that in aviation and to establish themselves with confidence in their chosen profession or avocation. This attitude, they believed, would pave the way for increased participation of women in aviation at a more equitable level. In fact, women's accomplishments in the 1970s did establish the broad base needed for the expanding presence of women in aviation that has subsequently unfolded.

9. Captains of Industry, Airlines, and the Military: 1980-1985

I believe all individuals should have the choice to pursue their talents and desires without limits. And I look forward to the day when there aren't any restrictions on what women can and cannot do.

Lt. Colleen Nevius, USN

At the midpoint of the 1980s, the United States has more women in aviation, who participate in a greater array of occupations, have access to more opportunities, and are protected by more laws, than does any other country in the world. It is revealing, however, that the ratio of women to men in aviation in the United States has not changed a great deal since 1940, although there are some signs that the percentage of women may be increasing. Bare statistics are not very illuminating in themselves, nor do they indicate the changing attitudes of American society. They do not explain the transformation in the circumstances under which women today build, pilot, or service airplanes, when contrasted with those who did so yesterday. The anecdotes related by women in aviation in the 1980s, both triumphant and painful, are still sharply reminiscent of the experiences of women from earlier periods. They are the same—heirs of historical patterns wrought decades earlier—and yet they are different—exemplars of new trends and new expectations.

Aviation does not offer women utopia in the 1980s, but there are factors that suggest they may be on the verge of a new era. The two most important indicators are American society's diminishing need to define women in aviation as a group and changing attitudes about the need to restrict women from combat. There is not unanimity of public opinion, but should these changes come to pass at some future date, the early 1980s will be hailed as the transition phase.

Professionally and socially women in aviation are becoming, for the most part, less isolated because of their gradually increasing numbers, and more confident because of their documented success. They have legal tools to reinforce their rights of participation. At the same time, airplane technology is becoming better adapted to female requirements. Human-factors engineers are designing cockpits that are more appropriate to women's smaller stature, just as they previously worked out a design for instrument lighting to accommodate the significant incidence of color blindness experienced by men.

The employment of women in aerospace continues to increase. In 1980 there were 253,900 women (21.4 percent) out of 1,185,000 employees in the field. In 1984 these numbers had grown to 280,600 women (23.4 percent) out of the total 1,197,200.[1] There were approximately 234,000 engineers and scientists working for aerospace companies in 1985.[2] Only a small percentage of this group were women, but they succeeded in attaining responsible positions and a high degree of recognition. Three of the five women honored in 1986 by The Aerospace Corporation, based in El Segundo, California, were engineers.[3] In 1985 Nancy Fitzroy was elected the first woman president of the American Society of Mechanical Engineers. She assumed office in June 1986.

There are a number of women who work at government agencies such as the Federal Aviation Administration and the National Aeronautics and Space Administration. The women astronauts, such as Sally Ride, Kathleen Sullivan, Anna Fisher, Rhea Seddon, and Judith Resnik have been, of course, the most visible federal employees. Many of the women astronauts are also pilots. All have had the opportunity to fly trainer aircraft. The history of the women in the space program is really a story apart from this volume; nonetheless, there seems to be an indelible connection for most people between these women and the concept of "women in aviation." In this way, the women astronauts serve as powerful role models for those women (and especially young girls) who would like to pursue a career in aviation or aerospace-related fields.

The changing attitudes and opportunities of individual women, be they glamorous astronauts or unknown engineers, have been reflected in American opinion. One obvious example of this change of opinion is a relatively new practice by public relations officers in various aerospace companies when developing photographic presentations for shareholders' annual meetings or corporate publications. The companies are now making serious attempts to include women and minority representatives. The women pictured may be tokens, but the images have helped women become more permanently integrated and accepted in a predominately male work force.

Small businesses continue to employ women as bush pilots, as agricultural aviation pilots, and in a host of jobs at small airports and fixed-base operations across the United States. Gayle Ranney, for example, works as a bush pilot in Alaska. When she was interviewed in 1981, there were about 400 female bush pilots in Alaska (out of 10,500). The job requires a certain brashness and sense of adventure, yet Ranney, who turned down jobs with commercial airlines, loves it because the flying is always challenging. Whether flying as air-taxi driver, volunteer coast guard, aerial geographer, or flight instructor, she finds each flight represents a totally new experience. She believes her greatest compliment is that no one has refused to fly with her simply because she happens to be a woman.[4]

Agricultural aviation is an area where women are both pilots and educators. Many states have agricultural aviation organizations for women who are in, or affiliated with, the profession. There are several women helicopter pilots who operate crop-dusting businesses.

There are an increasing number of women who are following the example of Olive Ann Beech, the premier woman entrepreneur in American aviation. Dr. Nydia Meyers, president and chief stockholder of the Al Meyers Airport Corporation in Tecumseh, Michigan, is an active pilot and a member of the Ninety-Nines, the Aircraft Owners and Pilots Association, and the Experimental Aircraft Association. Dr. Meyers has been equally successful in a career of scientific and medical research, publishing numerous papers and teaching college classes. Owners of a thriving airport in Tecumseh, both she and her husband, Al Meyers (who designed and produced three general aviation aircraft), have played a major role as general aviation advocates in the state of Michigan.[5]

More women are flight instructors than airport owners, however. Women can be found teaching at almost every airport in the United States. They are now somewhat taken for granted. An article about Morgan Greschel of Virginia, one of Janelle Aviation's two assistant flight instructors, highlighted her many accomplishments in general aviation. What the author saw as unusual about Greschel's accomplishments of 2,000 flight hours, an FAA Gold Seal Instructor award, and a variety of flying ratings, was not her sex but rather her youth. Greschel was only 25.[6]

Women serve in a multitude of capacities in the federal government. Elizabeth Dole, appointed U.S. Secretary of Transportation in 1983, was considered the most influential individual in American aviation. She had overall responsibility for federal aviation programs as well as those in the Coast Guard.

In 1983, about eight percent of all air traffic controllers were women. Many of these women belonged to an organization known as the Professional Women Controllers (PWC), which was founded in 1979. PWC originated as a dream of Jacqueline Smith and Sue Mostert when they met at the FAA Academy in 1968. There were many problems faced by women in this profession, principally because of their isolation. It was, and still remains, common for a female controller to be the first and only woman at a given location. Smith and Mostert wanted to create a women's organization for air traffic controllers as a network of communication that could break down the barriers of isolation. Thus, it was not created to compete with the union functions of the Professional Air Traffic Control Association. Naturally, PWC's membership was affected by the 1981 strike, which led to the disbanding of PATCO, but PWC was not forced to break up. The new union, the Air Traffic Control Association, is now an important supporter of PWC's activities, which include an active recruiting program to encourage women to enter the profession. The goal is to help women realize their full potential in the profession through education, communication, and support.[7]

Blanche Noyes was, for many years, the only woman to have permission to fly government aircraft. Today there are hundreds. These women may be FAA officials, or they may be involved in more exotic professions, such as aerial fire fighting. The first woman pilot with the Forest Service was Mary Barr, who later became an Air Safety Officer with the Forest Service's Fire and Aviation Safety management program. The second woman was Charlotte Larson. She has flown a variety of fire-fighting missions, which include aerial photography throughout the western part of the United States. Larson once received a letter of tribute from another woman pilot: "I don't know you, but I think I owe my job to you. The man who hired me was so impressed by your professionalism that he was convinced a woman could do this job."[8]

Professionalism is what ultimately convinces most people that women also make good commercial airline pilots. There are still only a few women who are working in this area. In 1985 there were 140 women pilots in the Air Line Pilots Association, representing less than half of one percent of the 33,500 total membership.[9] Despite the small numbers, a few women have earned the four stripes of an airline captain. Promotion for airline pilots, who begin as flight engineers, move up to first officer (copilot), and then to captain (pilot), is based on seniority. Women flying with presence, not simply because she was a woman but also because she was so young.

Most of the senior officers with Rippelmeyer's previous employers were veterans of World War II, old enough to be her father. Like most other women airline pilots, she learned to adapt to the generation difference with various coping mechanisms. She consciously cultivated a student-teacher relationship so as to reduce the tension that results among coworkers subjected to a significantly different workplace environment. She believed that it was her responsibility to make the captain and other male officers comfortable with the idea of a woman pilot, in as much as few had any experience or contact with professional women peers. Furthermore, from Rippelmeyer's perspective, the airlines as a whole seemed to do little to facilitate the process of integrating women and other minorities into their companies.

Age and experience are important factors in the process of integration, according to Rippelmeyer. She noted that at PEOPLExpress:

We're all the same age and most of the guys were hired after I was. I'm the senior person. I'm the captain. I'm in charge. Most of them, I'd say 90 percent of the men, have wives that work, so they're used to a professional woman and a professional woman's

FIGURE 62.—Charlotte Larson (left) became the first woman smoke jumper aircraft captain in 1983. Deanne Schulman (right) was the first woman to train and qualify as a smoke jumper, also in 1983. (Courtesy of USDA, Forest Service)

the older carriers say that it will take a long time before they can expect to reach the coveted "left-seat" position of captain.

Deregulation of airlines created new companies, and women pilots reaped the benefits of quick advancement. Such is the case of Lynn Rippelmeyer and Beverly Burns, both of PEOPLExpress (later to become part of Continental), who became the first women to be Boeing 747 captains. Rippelmeyer and Burns made their milestone flights—Rippelmeyer flew transatlantic and Burns flew transcontinental—on the same day. Each was unwilling to permit herself exclusive media honors, so they asked the company to schedule the flights on the same day.

An interview with Lynn Rippelmeyer revealed the high level of intelligence and professionalism striven for by all women pilots. She is bright, capable, and articulate, filled with observations about the change in societal attitudes toward women in the cockpit. In the late 1970s, when she first began working as a pilot for another airline, there was much resistance to her

FIGURE 63.—Captain Beverly Burns (left) and Captain Lynn Rippelmeyer (right) made aviation history when they captained two 747 aircraft on 19 July 1984. Burns was the first woman to captain a 747 cross-country, and Rippelmeyer the first woman to captain a 747 on a transatlantic flight. (Courtesy of David Hollander)

attitude, problems and view on life. I'm not the foreign species that I was to a lot of those other guys [at other airlines]. Even the military men have dealt with women peers. So it's not like somebody changed the rules on them and they were angry about it as I think some were a few years ago.[10]

There is one group of airline employees who are very supportive of women pilots—the flight attendants. Rippelmeyer, who first entered the airline world as a flight attendant, told an interviewer that, "As for the flight attendants I fly with, it's really touching and rewarding. They're real proud that we have a company that can create an environment that encourages a woman to be captain."[11]

The flight attendants of today are much different from their predecessors. The emphasis in training school is on safety, particularly the handling of potentially catastrophic situations such as accidents or terrorist attacks. The airlines want their passengers to be comfortable and have a sense of well-being, a traditional goal, but the companies no longer believe that only women with a certain appearance can cultivate that feeling. They want, in fact, to have as diverse a group of attendants as possible. There are about 60,000 attendants today, and 85 percent are women. A little more than half are married, and they

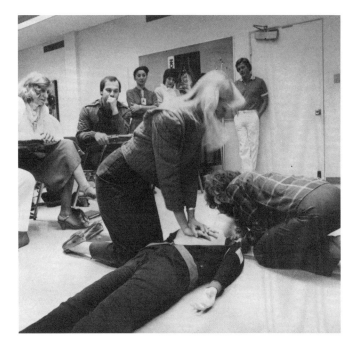

FIGURE 64.—Flight attendants practice cardio-pulminary resuscitation (CPR) techniques as a part of their recurrent training program. First aid, survival tactics, and terrorist handling are all skills taught to flight attendants. (Courtesy of the Association of Flight Attendants, S.I. photo 86-11860)

range in age from 19 to 60. A range of acceptable heights is still prescribed, but the only restriction on weight is that it be in proportion to height.[12] Educational requirements have increased, and some airlines now require a college diploma.

One of the major differences between today's attendants and the hostesses and stewardesses of the past is salary. The increased recognition and status of the attendants is reflected in a proper professional salary scale. These wages, plus the many benefits, are not guaranteed, however. The status of the flight attendants today is directly linked to the power and strength of the unions to which they belong.

Some airline executives in the 1980s are inclined to ask the various employee groups to bear the brunt of cutbacks in an age of deregulation. The flight attendants have been particularly vulnerable because their status is so newly acquired. Also, it would seem that because most of their membership is female, their unions continue to have a disproportionate struggle. The appearance of company advertisements in newspapers for new attendants is often a sign of impending contract negotiations. Essentially, a company is attempting to avoid the threat of a strike by hiring a corps of immediate (albeit temporary) replacements. For many flight attendants, the union structure seems to provide some protection against such measures and some assurance of progress and security in an uncertain future.[13]

The military does not experience the same financial constraints and market competition as the airlines. The mission of national defense overrides such issues. In recent years, the passage of civil rights laws and the advent of the all-volunteer force have reshaped many traditional military ideals and standards, particularly regarding women in the military. The central question today is whether or not women should serve in combat roles, an issue over which Americans remain deeply divided. Military planners and strategists are concerned that women would adversely affect troop cohesion and thereby diminish combat effectiveness. The Army has begun to commission studies on these topics through the Army Research Institute for the Behavioral Sciences. Report and test results have yet to generate either unified interpretation or new policy standards.

Some politicians actively support full involvement, including a draft for men and women. Others believe women should be limited to support functions. It is interesting that few, however, suggest removing women altogether. The predominant view in the 1980s, as in the past, supports only a noncombat role for women, although the participation of women in the military has expanded significantly.

Women's organizations are generally more supportive of the idea of women serving in combat. Still, large feminist organizations such as NOW are in a quandry. They ardently support an expanded role for women in the military, as a matter of principle, but some members believe that by supporting women in the military (including those in aviation) they are helping perpetrate and strengthen an institution they would like to see eliminated, or at least drastically curtailed. Others argue for "equal rights, equal risks." Support for this position comes from groups such as DACOWITS, the Women's Military Pilots Association, and the Society of Women in Military Aviation. They actively promote a program of military professionalism for women.

Female military pilots still account for less than one percent of the total number of pilots in the armed services. However, women have been participating since the mid-1970s, necessitating an effort by the Department of Defense to develop reasonable career options for women in military aviation. This has resulted in refining previous definitions of combat. Nonetheless, it is only with great determination that women in military aviation achieve a challenging program of career development. Women fly reconnaissance and cargo airplanes that ferry supplies into combat areas. They fly air refueling tankers as well as attend the Navy's Test Pilot School at Patuxent River, Maryland. They are carrier qualified (meaning that they can take off and land on aircraft carriers) and have flown in combat-scenario flight exercises. Three aircraft maintenance teams that are part of the Rapid Deployment Force are headed by women.[14] The first all-women crew completed a Military Airlift Command mission flight on 9 May 1983 from McGuire Air Force Base to Rhein-Main Air Force Base, West Germany. The crew attracted media attention because all members were female. While there was nothing unusual about the way they performed their jobs, the publicity demonstrated to other women in the Air Force and to the public that women can accomplish such tasks without threatening national security.[15]

Every significant military action in recent years—including Grenada and Lebanon—has involved women pilots flying as part of the mission.[16] Military jobs are no longer simply classified as "combat" or "noncombat"; they are often graded on a scale of combat likelihood. Lower-risk positions have been opened to women. This means that women may be assigned to regions like the Demilitarized Zone in South Korea. Many women in the military hope that opening up these jobs will permit greater numbers of women a chance to achieve their best and enhance their chances for promotion. It also increases the possibility that "women [will be] brought home in body bags."[17] Until

FIGURE 65.—Ensign Mary A. Crawford became the U.S. Navy's first woman Naval Flight Officer when she graduated from the Interservice Undergraduate Navigator Training program in June 1981. (Courtesy of the U.S. Air Force, DoD Still Media Depository, DN-SN-82-03497)

such time as combat restrictions are lifted, however, the total number of women in military aviation will continue to be small.

Women are very active in general aviation. In 1984, 44,409 women were licensed pilots in the United States, representing a little more than six percent of the total 720,534 active pilots.[18] Women are involved in every aspect of general aviation from hang gliding to human-powered flight. Recently women have built and exhibited aircraft shown at the Experimental Aircraft Association's annual Oshkosh Fly-In.[19] One woman serves on the ground crew of the Goodyear blimp *Columbia.* Women still fly competitively. Some fly or assist teams attempting new FAI world records. Brooke Knapp and Jeanna Yeager are two examples. Others take part in air races. The All Women's International Air Race (the Angel Derby) still exists. The major event of recent years is the Precision Flying Competition. In November 1981, Janet Green, president of the Ninety-Nines, signed the resolution that gave the Ninety-Nines sole responsibility for sponsorship, training, and organization of the U.S. Precision Flight Team, as well as management of the regional and national meets. In 1985 the Ninety-Nines hosted the sixth World Precision Flight Team Championship at Kissimmee, Florida.

Precision flying is not the only activity in which the

FIGURE 66.—Lieutenant Colleen Nevius, USN, was the first woman to graduate from the Naval Test Pilot School in Patuxent River, Maryland, in June 1983. Lt. Nevius is shown in connection with tests she conducted on the first modified HH-46A Seaknight helicopter. (Courtesy of Lt. Colleen Nevius, USN, S.I. photo 86-12200)

Ninety-Nines are involved. They continue to support the National Intercollegiate Flying Association, but their work in aerospace education is perhaps their most important project. Former international president Hazel Jones has constantly advocated this effort both internally and externally. Jones believes that an understanding of the past coupled with an awareness of new pioneers are the twin sources of inspiration for future success. It is not a coincidence that while the organization's membership is only about 6,500, many of the women who are in positions of responsibility and power in the aviation world are also Ninety-Nines; for example, Dr. Dora Strother, Manager of Human Factors Engineering and Cockpit Arrangement at Bell Helicopter Textron; Carol Rayburn, Manager of the FAA's General Aviation and Commercial Divisions; and

FIGURES 67-70 (facing page).—Expanded opportunities for women in military aviation: 67 (top left), AMN Maudell P. Thompson is a Titan Missile facilities technician for the 533rd Strategic Missile Squadron at McConnell Air Force Base, Kansas. (Courtesy of the U.S. Air Force, DoD Still Media Depository, DF-ST-82-04427) 68 (top right), Captain Barbara Akins served as copilot during the landing of a C-141 MEDEVAC aircraft at Lajes Field, Azores. She was part of the first all-female flight crew to fly an overseas mission of a C-141 from McGuire Air Force Base, New Jersey, to Rhein-Main Air Base, West Germany. (Courtesy of U.S. Air Force, DoD Still Media Depository, DF-ST-84-01749) 69 (bottom left), Sergeant Munn hangs a wet parachute in the chute tower of the Survival Equipment and Accessories Branch. She is a member of the 52nd Component Repair Squadron at Spangdhlem Air Base, West Germany. (Courtesy of the U.S. Air Force, DoD Still Media Depository, DF-SN-83-03220) 70 (bottom right), a ground crew chief inspects the intake of an F-111A aircraft. (Courtesy of the U.S. Air Force, DoD Still Media Depository, DF-ST-82-08323)

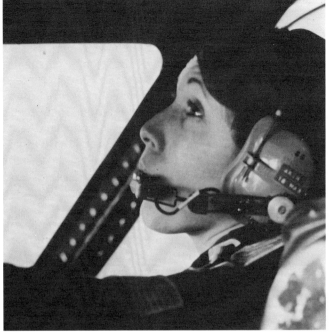

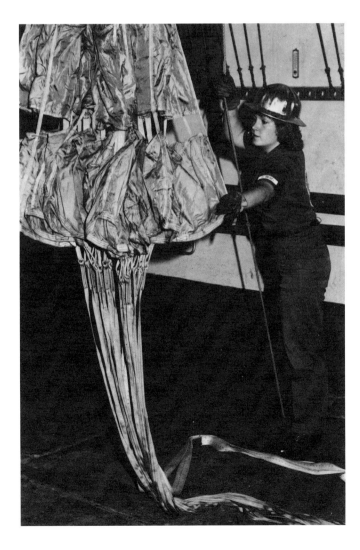

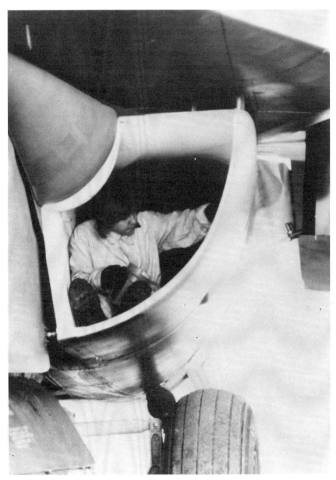

Jean Ross Howard, Director of Helicopter Activities, Aerospace Industries Association and Whirly-Girls Executive Director.

The present role of United States women in aviation may be considered a part of the legacy of Amelia Earhart. Despite her disappearance in 1937, her image continues to be the sine qua non for the history of women in aviation—always an example that elicits admiration and imitation. As a flyer, Earhart embodied the essence of the "old guard," those grand, individualistic women of aviation from the early pages of flight history. As a feminist, she was (in her words) before her time, but her example of commitment to performance as the chief measurement of success has proved to be the key that opened the door for other women who loved flying. Individual after individual won an air race, broke a speed record, headed an aircraft company, or organized a flight attendants' union. Collectively their importance amounted to more than a simple sum of their accomplishments. By bulk of numbers and also by the power of networking organizations, they demonstrated that women are a legitimate part of the aviation scene. They did this so well that they succeeded in freeing themselves from the confines of the female ghetto—the separate women's unit. The strength of the group enabled the individual to move beyond the confines of the group and to be integrated into the mainstream of aviation personnel. Nowhere has this been more dramatic than in the United States armed forces.

United States women in aviation have both shaped and been shaped by many of the social issues and technological developments that have stirred the country in the 45 years covered by this study. They now know that they have the aptitude, the access to education, the legal standing, and the professionalism that will enable them to walk through the recently opened doors of opportunity into a new era of full participation in aviation, not only in the United States, but around the world.

Appendixes

I. North American Women Who Served in Great Britain with the ATA during World War II

The source for this list is W.A.S.P., *Women Airforce Service Pilots, WWII, 1982 Roster*, p. 4. Ann Wood, a participant, verified the fact that all these women listed flew. In addition, Polly Potter was picked by Cochran and sent to Britain but she did not pass the medical exam.

UNITED STATES WOMEN PILOTS RECRUITED BY COCHRAN

Myrtle Rita Allen (Carter)
Opal Pearl Anderson (Averitt)
Dorothy Rita Bragg (Hewitt)
Emily Chapin
Catherine Del Van Doozer
Virginia Farr
Mary Estelle Hooper Ford
Suzanne Humphreys Ford (de Florez)
Virginia Garst
Una Goodwin
Evelyn Hudson (Richards)
Margaret Elizabeth Lennox (Drown)
Nancy Jane Miller (Livingston)
Mary Webb Nicholson
Winnie Rawson Pierce (Beasley)
Hazel Raines
Helen Richey
Roberta Sandoz (Leveauz)
Louise E.M. Schuurman
Edith Foltz Stearns
Grace Stevenson
Ann Watson Wood (Kelly)

CANADIAN WOMEN PILOTS RECRUITED BY COCHRAN

Helen Harrison
Gloria Large

OTHER UNITED STATES WOMEN PILOTS WHO JOINED ATA ON THEIR OWN

Sheila Garrett
Ruby D. Garrett
Violet Beatrice Milstead
Leslie Carnes Murray
Joan Ratcliff
Jane Graham Plant Spencer

II. Summary of Information Provided for Congressional Inquiry into the WASP, 1944

The source of this summary is Tunner, memorandum, 10 April 1944. Nancy Love compiled the survey for Colonel Tunner, who sent it on to General Arnold.

Questions and Statements in Congressional Inquiry	*Survey Response*
1. Is there resentment among male pilots?	Resentment does exist amongst a minority of the Ferrying Division male pilots against the WASPs. This minority is, however, quite virulent. It is not believed that resentment exists among the supervisory or operational personnel whose sole interest is the efficient performance of the ferrying mission. These personnel are familiar with the pilot shortage in the Ferrying Division and appreciate the WASP's contribution.
2. Favoritism is being practiced in giving WASPs more opportunity to fly than the male pilots.	The statement is incorrect in the Ferrying Division. WASPs' flying time to date has consistently averaged less than flying time of male pilots.
3. Do WASPs take their turn as to frequency and type of assignments? If so, how is this controlled?	WASP pilots take their turn as to frequency and type of assignment. WASPs are formed into squadrons just as are our men pilots. Assignments to ferrying duty are awarded to squadrons consistent with their daily pilot strength and a careful survey reveals that there is not any partiality shown towards WASPs in providing them with greater frequency in delivery. Normally, any pilot in the Ferrying Division, including the WASPs, who has returned to his base is placed in the bottom of a roster consistent with the type of aircraft he is qualified to fly, and ferrying missions are assigned to the individual who has been longest at base.
4. Male pilots sit on the ground for days while WASPs get assignments and are kept busy.	This statement is inaccurate. All ferrying pilots have been used without anyone waiting an undue length of time for assignment. For several months past, the Ferrying Division has been extremely busy in its domestic operations, and all pilots, regardless of sex, have been utilized without anyone waiting an undue length of time for assignment. It is to be noted, however, that frequently some pilots are qualified only on a few types of aircraft and, due to a shortage of transition planes to qualify them on advanced types and due to natural inability to advance rapidly, must sit on the ground while other pilots who are well qualified on many types are of much more value in the ferrying mission and hence fly more types of aircraft. This is true amongst the WASPs just as it is amongst some male pilots.

III. Tables

TABLE 1.—Occupations open to women in the WAC as well as the number of women serving as of 31 January 1945; women trained as aircraft mechanics and radio operators entered service in 1944.

Specialty	Number of women	Percent of total female personnel
Administrative	14,011	46.0
Duty	5,726	18.8
Medical	1,879	6.2
Vehicle Operator	1,628	5.3
Supply	1,364	4.5
Photographic	972	3.2
Aviation Specialist	656	2.1
Wire Communications	650	2.1
Airplane Maintenance	617	2.0
Radio Mechanic	359	1.2
Radio Operator	301	1.0
Weather	242	0.8
Utility Construction	169	0.6
Auto Mechanic	82	0.3
Armament	51	0.2
Air Crew	20	0.1
Radar Operator	5	<0.05
Misc.	1,698	5.6
Total	30,430	100.0

Source: Treadwell, *The Women's Army Corps,* p. 720 [quotes data extracted from "Strength of AAF Personnel in WAC within Continental U.S. by Command, SSN, and RACE." SC-PS-123, AAF Office of Statistical Control].

TABLE 2.—WASP assignments at demobilization.

Training Command	620
Air Transport Command	141
Second Air Force	80
Fourth Air Force	37
First Air Force	16
Weather Wing	11
Proving Ground	6
Air Technical Service	3
Headquarters AAF	1
Troop Carrier	1
Total	916

Source: Cochran, "Final Report," p. 28.

TABLE 3.—Cost of training each WASP pilot.

Tuition, student salary, airplane depreciation		$6,265.35
Additional costs		
Maintenance cost, material, labor, gas and oil	$3,023.50	
Personnel, military and civilian	540.10	
Equipment	89.56	
Travel	18.00	
Uniform	326.06	
Medical examination and hospitalization	66.59	
Communications	8.80	
Amortization, crash truck, link trainer, vehicles	95.66	
Maintenance, administrative vehicles	13.64	
Adjustment for eliminees	1,703.44	
Total additional costs		5,885.35
Total cost per graduate		$12,150.70

Source: U.S. Congress, House, Committee on the Civil Service, *Concerning Inquiries.*

TABLE 4.—Active pilot certificates issued by the United States government; information about women pilots was not collected for 1946-1959; total number of women's certificates in some years does not equal the sum of the preceding columns because some certificates do not fall in these categories or because some women held more than one certificate (dash = information not available, na = not applicable).

Year	Women						Men + Women
	Student	Private	Commercial	Airline transport	Nonpilot	Total	Total
1940	311	476	86	–	na	902	31,264
1941	202	1,803	113	–	na	2,145	63,113
1942	–	3,009	184	–	na	3,206	100,787
1943	–	3,739	373	–	na	4,112	124,050
1944	–	4,211	618	–	na	4,829	132,435
1945	–	4,166	956	–	na	5,112	141,280
1960	5,748	3,425	738	25	2,080	9,966	348,062
1961	5,875	3,554	733	26	2,084	10,246	352,860
1962	5,939	3,683	728	27	2,116	10,512	365,971
1963	6,755	4,004	813	51	2,142	11,757	378,700
1964	8,179	5,218	1,047	55	2,235	14,627	431,041
1965	10,106	6,147	1,137	40	2,280	17,555	479,770
1966	11,424	7,319	1,317	61	2,376	20,265	548,757
1967	13,173	8,775	1,479	57	2,477	23,659	617,931
1968	16,243	10,164	1,691	71	2,707	28,401	691,695
1969	16,055	11,174	1,824	76	2,912	29,415	720,028
1970	15,787	11,409	1,897	79	3,078	29,472	732,729
1971	16,417	12,332	2,032	88	3,413	31,216	741,009
1972	17,053	13,391	2,196	101	3,594	33,001	750,869
1973	18,593	13,232	2,083	95	3,074	34,356	714,607*
1974	19,298	14,465	2,596	116	3,471	36,943	733,728
1975	19,600	14,952	2,733	137	3,809	37,934	728,187
1976	22,254	15,838	2,857	160	4,252	41,643	744,246
1977	25,705	17,702	3,090	193	3,672	47,294	783,932
1978	26,354	19,267	3,306	270	3,984	49,874	798,833
1979	26,714	20,275	3,618	361	4,350	51,733	814,667
1980	26,006	21,554	3,993	480	4,779	52,902	827,071
1981	22,591	19,602	4,101	584	5,201	47,721	764,182
1982	19,958	19,388	4,257	749	5,697	45,305	733,255
1983	18,696	18,801	4,281	884	6,151	43,648	718,004
1984	19,435	18,616	4,232	1,032	6,591	44,339	722,376
1985	19,058	17,974	4,185	1,184	6,017	43,483	709,540

*Purge of 26,000 duplicate certificates

Sources:

1940-1945 Aeronautical Chamber of Commerce of America, *Aircraft Yearbook, 1940-1946*

1960-1969 U.S. Department of Transportation, *U.S. Civil Airmen Statistics*, June 1971. Nonpilot certificates include mechanics, parachute riggers, dispatchers, control tower operators, and ground instructors.

1970-1979 Ibid., 1979-1985/86. Nonpilot certificates include mechanics, parachute riggers, ground instructors, dispatchers, and control tower operators.

1980-1985 Ibid., 1985/86. Total pilot certificates include student, private, commercial, airline transport, helicopter only, glider only, and lighter-than-air only categories. Nonpilot certificates include mechanics, parachute riggers, ground instructors, dispatchers, flight engineers, and flight navigators.

Notes

Full bibliographic citations will be found in "References."

1. Students and Teachers, Clubs and Colleges
(pages 3-11)

Epigraph: from unidentified contemporary newspaper clipping.

[1]The Ninety-Nines, Inc., *The History of the Ninety-Nines, Inc.*, p. 29.

[2]Link trainers were special training units manufactured by the Link company; they were used to teach pilots how to fly exclusively by instruments. The trainers simulated the cockpit, as well as all the physical sensations of flight. Instructors created the flight conditions, monitored their students' flights, and taught the skills needed to navigate. It was a rigorous occupation that required the ability to pilot an aircraft. Women who worked as instructors "knew" how to fly military aircraft, even if they never had the opportunity to pilot a real vehicle.

[3]The Ninety-Nines, Inc., *The History of the Ninety-Nines, Inc.*, p. 10.

[4]Arthur, "Airways to Earning," pp. 34-35.

[5]The Ninety Nines, Inc., *The History of the Ninety-Nines, Inc.*, p. 23; "Babies, Just Babies," np.

[6]Planck, *Women with Wings*, p. 248; Arthur, "Now You Can Learn to Fly," pp. 321, 336; Arthur, "Wings for the Working Girl," p. 41.

[7]Opal Kunz was also a founder of the Betsy Ross Corps, which was started in 1931. It was a well-trained, women's paramilitary air corps dedicated to the sole purpose of national defense, through humanitarian relief work. (The Ninety-Nines, Inc., *The History of the Ninety-Nines, Inc.*, p. 11.)

[8]Arthur, "Now You Can Learn to Fly," pp. 320-321, 336.

[9]Ibid., p. 336.

[10]The original goal of the WFA was to provide a "trained, disciplined corps of women to replace men behind the air lines in a national emergency." (Arthur, "Now You Can Learn to Fly," p. 336.) A member of the Washington, D.C., chapter was quoted as saying: "Every member of the Chapter is eager to serve in the women's Auxiliary Ferry Service and the Group hopes to help as many students as possible accumulate flying time to qualify for this service." ("D.C. Women Flyer's Chapter," np.)

[11]Arthur, "Wings for the Working Girl," p. 136; Arthur, "Now You Can Learn to Fly," p. 336.

[12]Wixson, "Air Ferrying Service," np.

[13]Laura Brown, "Flier or No," p. 8; Logan, "Women Volunteers," np.

[14]Shoemaker, "Air Schooling for Milady," pp. 39-40, 42, 112; Planck, *Women with Wings*, p. 225.

[15]Pisano, "A Brief History," pp. 21-25, 27; Planck, *Women with Wings*, pp. 139-143.

[16]Arthur, "Airways to Earnings," p. 55.

[17]Planck, *Women with Wings*, p. 140. The CAA paid a standard fee of $290 for each student to the flight operators and a lump sum of $200 to each institution giving the ground school course. ("The Civilian Pilot Training Program," p. 4.) Rough estimates of actual expenses based on figures supplied by the CAA indicate that the costs were $390 per student in fiscal 1940 and $640 per student in fiscal 1941 (the fiscal year ran from 1 July to 30 June). (U.S. Civil Aeronautics Authority, *Wartime History of the Civil Aeronautics Administration*, p. 23.)

[18]Howard, interview.

[19]Planck, *Women with Wings*, pp. 150-151.

[20]Ibid., p. 150.

[21]Strickland, *The Putt-Putt Air Force*, p. 56.

[22]Tubbs, letter.

[23]"Women Instructors Graduated," p. 239; Slack, "Tennessee's Airwomen," p. 47.

[24]Oakes, *United States Women in Aviation: 1930-1939*, p. 9.

[25]CAA Administrator C.I. Stanton stated: "It is my opinion that since women have always excelled in instructing and have done most of the teaching of our nation, this should be their natural function in aviation. Our problem is to give the 1,000,000 boys who will graduate into the draft each year, flight training. I believe we should train at least 200,000 of them each year. To do that, we shall need at least 5,000 women instructors." (Slack, "Tennessee's Air Women," p. 46.)

[26]Ibid., p. 128.

[27]Ibid., p. 128; Knapp, *New Wings for Women*, p. 166.

[28]"Women Instructors Graduated," p. 239; Chapelle, *Girls at Work in Aviation*, p. 44.

[29]Planck, *Women with Wings*, pp. 154-155.

[30]Bates, "Lady with Wings," p. 55.

[31]Chapelle, *Girls at Work in Aviation*, p. 40.

[32]Mary Steele Ross, *American Women in Uniform*, pp. 49-51; Nichols, *Wings for Life*, pp. 267-278.

[33]They included C.R. Smith, President of American Airlines; William Patterson, President of United Airlines; Jack Frye, President of TWA; Eddie Rickenbacker, President of Eastern Airlines; C.V. Whitney, Chairman of the Board of Pan American Airways; and Walter and Olive Ann Beech, owners of Beechcraft.

[34]"Now a Civil Air Patrol," p. 44.

[35]Mary Steele Ross, *American Women in Uniform*, p. 20.

[36]Reiss, "Ground Pilot," pp. 122, 126; Charles P. May, *Women in Aeronautics*, p. 165.

[37]Reiss, "Ground Pilot," pp. 122, 126.

[38]Chapelle, *Needed*, p. 22.

[39]"Lieut. Willa Brown," np.; Charles P. May, *Women in Aeronautics*, p. 165.

[40]Chapelle, *Needed*, p. 19.

[41]Knapp, *New Wings for Women*, p. 35. Initially these scholarships, which were offered at the Massachusetts Institute of Technology, New York University, California Institute of Technology, University of Chicago, and the University of California at Los

Angeles, were available only to men, but in March 1942 women were included in the program.

[42]Knapp, *New Wings for Women*, pp. 35-38, 43. An interesting side note is that the CAA asked Tonkin to conduct a recruiting tour to interest women in applying. She was not very successful and upon returning to Washington she commented: "Women usually consider calculus and physics rather dry and colorless subjects and shy away from them in college. So I found few with such training plus a private pilot's license." (Ibid., p. 38.)

[43]Edward J. Gardner, "Help Wanted!" pp. 48, 128.

[44]Oakes, *United States Women in Aviation: 1930-1939*, p. 9; Noyes, "Air Marking—Reversed," p. 107; "The Reminiscences of Blanche Noyes," np.

[45]Glen Gilbert, *Air Traffic Control*, p. 11.

[46]After the war, many military-trained women would transfer their skills to the CAA.

2. Coffee, Grease, Blueprints, and Rivets
(pages 12-26)

Epigraph: "Demand Growing," pp. 102, 112.

[1]Planck, *Women with Wings*, pp. 196-197; Chapelle, *Girls at Work in Aviation*, pp. 198, 201.

[2]Nielsen, *From Sky Girl to Flight Attendant: Women and the Making of a Union*, p. 37.

[3]Ibid., p. xvii.

[4]Ibid.

[5]"Coffee, Tea or Customer Service Manager?" p. D5.

[6]Planck, *Women with Wings*, p. 197; Knapp, *New Wings for Women*, pp. 76-77.

[7]Nielsen, *From Sky Girl to Flight Attendant: Women and the Making of a Union*, p. 17.

[8]Knapp, *New Wings for Women*, pp. 73, 80.

[9]Nielsen, *From Sky Girl to Flight Attendant: Women and the Making of a Union*, p. 24.

[10]Ibid., pp. 15, 25; Aircraft Industries Association of America, Inc., *Aviation Facts and Figures 1945*, p. 28 [quotes Department of Labor, Bureau of Labor Statistics, "Wartime Development of the Aircraft Industry," *Bulletin* 800, November 20, 1944, p. 20]; Planck, *Women with Wings*, p. 195.

[11]Nielsen, *From Sky Girl to Flight Attendant: Women and the Making of a Union*, p. 23; Barger, *The Transportation Industries 1889-1946*, p. 266 [quotes from *Civil Aeronautics Administration, Statistical Handbook of Civil Aviation*, 1948 issue].

[12]It was the no-marriage rule which absolutely distinguished flight attendants from women in other wartime professions. It was in World War II that for the first time married women outnumbered single women in the work force. In particular Karen Anderson noted that (1) wives of servicemen were three times more likely to work than those women whose husbands were at home, and (2) women over 35 or between the ages of 14 and 25 composed most of the female work force. (Anderson, *Wartime Women*, pp. 4-5.)

[13]This observation was expressed by John R. Hill, Santa Barbara, California, 7 March 1981. (Nielsen, *From Sky Girl to Flight Attendant, Women and the Making of a Union*, p. 30.)

[14]Chapelle, *Girls at Work in Aviation*, pp. 196-197.

[15]Planck, *Women with Wings*, p. 191.

[16]Hays, "The WAMS," pp. 38-39, 168; Chapelle, *Girls at Work in Aviation*, p. 79.

[17]"The Month," (December 1942) pp. 10-11. 191 women were assigned to the famous Clipper ships.

[18]"Fleet—'Feminine Army,'" p. 27.

[19]"Transatlantic—'Respect for Women,'" p. 15. It is interesting that the article emphasized Betty Travis' background in automobile engine repair and the fact that she could "throw a baseball like a man." There was an implied connection made by the article in highlighting these traditionally masculine pursuits.

[20]"Transatlantic—The Women,'" pp. 12-13.

[21]Chapelle, *Girls at Work in Aviation*, pp. 181-184. Women served in in other non-production-line positions such as administrative assistants, personnel officers, statistical assistants, and secretaries. See "The Girls Everyone Likes," pp. 6-7, 21, and "The Forgotten Woman," pp. 6-7, 21, for further information on the experiences of women in these occupations.

[22]Eaves, "Wanted," p. 133.

[23]Rossiter, *Women Scientists in America*, pp. 91, 173, 226, 389; Hacker, "Mathematization of Engineering," pp. 45-46; Eaves, "Wanted," p. 158.

[24]*Women in Aircraft Engineering*, pp. [3, 5-7, 11].

[25]Ibid., pp. [5-7]; Erler, letter; *Women in Aircraft Engineering*, p. [7].

[26]Ahnstrom, "Look . . . Women," p. 31; Chapelle, *Girls at Work in Aviation*, pp. 153-154.

[27]Ahnstrom, "Look . . . Women," p. 66.

[28]Chapelle, *Girls at Work in Aviation*, p. 154; Chapelle, *Needed*, p. 169.

[29]Hacker, "Mathematization of Engineering," pp. 48-49; Baker, *Wanted*, p. 45; Monroe, "Women Artists Are Different," p. 4.

[30]Planck, *Women with Wings*, pp. 219-220.

[31]Knapp, *New Wings for Women*, pp. 169-179; Chapelle, *Girls at Work in Aviation*, pp. 151-153.

[32]Knapp, *New Wings for Women*, p. 179.

[33]Eaves, "Wanted," p. 159.

[34]Bender, telephone interview.

[35]"Hellcat Teasers," p. 123.

[36]Roberts, "You Can't Keep Them Down," p. 91; Knapp, *New Wings for Women*, pp. 11-12; "Hellcat Teasers," p. 123.

[37]Roberts, "You Can't Keep Them Down," p. 91.

[38]Ibid.

[39]Bright, *The Jet Makers*, pp. 2-3; Jablonski, *America in the Air War*, p. 18.

[40]Jablonski, *America in the Air War*, p. 28.

[41]Aircraft Industries Association of America, Inc, *Aviation Facts and Figures 1945*, pp. 20-21, 24, 35 [pp. 20-21 quote Department of Labor, Bureau of Labor Statistics, "Wartime Development of the Aircraft Industry," *Bulletin* 800, Nov. 20, 1944, p. 5; p. 24 quotes the same Department of Labor, Bureau of Statistics, *Bulletin*, p. 8].

[42]"Women Are Welcome," p. 22.

[43]Aircraft Industries Association, Inc., *Aviation Facts and Figures 1945*, pp. 20-21, 24 [see note 41, this chapter, for quoted sources].

[44]Anderson, *Wartime Women*, pp. 36-40.

[45]"Demand Growing," p. 112.

[46]"The Gals Come Through," p. 7.

[47]Ibid., p. 12.

[48]Anthony J. Smith, "Menstruation and Industrial Efficiency," pp. 1-5.

[49]Neville, "Education Alone," p. 89. For a contrasting account, read Chapelle, *Girls at Work in Aviation*, p. 90.

[50]"Demand Growing," p. 112; Baker, *Wanted*, p. 72; "The Gals Come Through," p. 7; "It's Up to the Women," p. 6. It is important to note that this idea is somewhat misleading. Labor history indicates that many women, including some married women with families, were in the workplace during the 1930s. Nearly 25 percent of the adult female population (21.2 million females over age 14) were gainfully employed during the decade prior to the war (according to the "General Report of Occupations," *Fifteenth Census of the United States 1930*, vol. V, p. 272). Proportionally twice as many black women as white women were in the work force. The basic fact was that most working-class and many middle-class men "were not paid enough to support their families according to the American standard of living." Thus it is clear that the trend of increasing numbers of women in the workplace, though small and strongly resisted by married white women, was well established prior to the war. (Wandersee, *Women's Work*, p. 1.)

[51]"Alaska," p. 19.

[52]Common in the aircraft-production plants during World War II was the following story. It was often used by recruiters as a "parable" to illustrate the importance of attitude and patriotism.

"Three girls were inspectors of valve spring washers in an airplane engine factory. They worked side by side, each doing exactly the same job as the others. A visitor passing through the factory asked one of them what she was doing.

"She smiled and said, 'I'm making eighty-five cents an hour,' and went back to her work.

"The visitor asked the second one what she was doing.

"'I'm inspecting valve spring washers for bomber engines,' she said seriously.

"When the visitor repeated his question to the third girl, she looked up happily, without letting her fingers miss a motion.

"'I'm helping to win the war, sir,' she said." (Chapelle, *Needed*, pp. 149-150.)

[53]Baker, *Wanted*, p. 11.

[54]Ibid., p. 191.

[55]Thorburn and Thorburn, *No Tumult, No Shouting*, p. 66.

[56]Bowman, *Slacks and Callouses*, p. 2. The swing shift ran from 4:30 pm to 1:00 am and paid an additional 8 cents per hour.

[57]Ibid., p. 165.

[58]Ibid., p. 168.

[59]Bernardin, *Women in the Work Force*, p. 31. Chapelle's books and similar ones by other authors were part of a carefully orchestrated propaganda program designed to convince women that working in aircraft factories was both fitting and proper. They were written for teenaged girls and contain many mini-biographies of women as well as specific instructions on how to enter the field.

[60]Chapelle, *Girls at Work in Aviation*, p. 7.

[61]Bernardin, *Women in the Work Force*, p. 31; Baker, *Wanted*, p. 23; Planck, *Women with Wings*, pp. 209-210.

[62]"The Gals Come Through," p. 7. Retooling the factories for wartime production marked a dramatic change for the manufacturers. The United States government enlisted the aid of automobile manufacturers to help aircraft companies change from a "job shop" system to line production. (Bright, *The Jet Makers*, p. 4.) It is interesting that the two new factors—women and production line techniques—just as with the case of women and absenteeism, are linked together.

[63]"It's Up to the Women," p. 7.

[64]Despite claims to the contrary, women did earn less. It was during this period that the National War Labor Board (NWLB) created two important policies to counter the problems of pay discrimination. Title II of Executive Order 9250 was a Wage and Salary Stabilization Policy, which promoted wage adjustments for the "correction of maladjustments or inequalities, the elimination of substandards of living and the correction of gross inequities." The second policy was General Order No. 16, which stated "that wages for women could be increased without approval of the NWLB to 'equalize the wage or salary rate paid to females with rates paid to males for comparable quality and quantity of work on the same or similar operations.'" (Beatty and Beatty, "Job Evaluation and Discrimination," p. 211; Aircraft Industries Association, Inc., *Aviation Facts and Figures 1945*, p. 30 [quotes Department of Labor, Bureau of Labor Statistics, "Hourly Earnings in Aircraft Engine Plants, August, 1943," Serial No. R1632, pp. 5, 6; also cites Department of Labor, Bureau of Labor Statistics, "Wages in Aircraft Propeller Industry, October 1942," Serial R1526, p. 8]).

[65]Corn, *The Winged Gospel*, particularly Chapter 3 and pp. 126-127; Baker, *Wanted*, p. 173; Chapelle, *Needed*, p. 215.

[66]"Madame Mechanic," p. 136.

[67]Anderson, *Wartime Women*, pp. 7-8.

3. Daughters of Minerva
(pages 27-43)

Epigraph: Unidentified newspaper clipping of a weekly column, "Just Folks," by Edgar Guest.

[1]De Pauw, "Women in Combat," p. 210; see also Holm, *Women in the Military*, pp. 3-15.

[2]Hancock, *Lady in the Navy*, pp. 22-23.

[3]Thaden, *High, Wide and Frightened*, pp. 229-254.

[4]Holm, *Women in the Military*, pp. 18-20; Keil, *Those Wonderful Women*, p. 100; U.S. Congress, House, Select Subcommittee of the Committee on Veteran's Affairs, *To Provide Recognition*, pp. 50-51.

[5]Jacqueline Cochran's September 1939 record was 305.926 mph for 100 km, Burbank to San Francisco in a Seversky. (Planck, *Women with Wings*, p. 318.

[6]Craven and Cate, *The Army Air Forces*, p. 528; Backus, *Letters From Amelia*, pp. 149, 171.

[7]Cochran and Odlum, *The Stars at Noon*, p. 6.

[8]Ibid., p. 40.

[9]The record was from Burbank, California, to Cleveland, Ohio, 2,042 miles in 8 hours, 10 minutes, and 31 seconds. (Cochran and Odlum, *The Stars at Noon*, p. 65).

[10]Craven and Cate, *The Army Air Forces*, p. 528 [quotes USAF Historical Studies, No. 55: "Women Pilots with the AAF, 1941-1944," pp. 2-3, and H.H. Arnold, *Global Mission* (New York, 1940), p. 311].

[11]Keil, *Those Wonderful Women*, p. 48.

[12]Margaret C. Love, interview. Air marking involved painting large markings (in all the major American cities) that were visible from an aircraft as a navigational aid to pilots. Love's responsibility was for the East Coast region, New York State in particular. (Knapp, *New Wings for Women*, p. 52.)

[13]Knapp, *New Wings for Women*, pp. 52-53; Oakes, *United States Women in Aviation: 1930-1939*, p. 62.

[14]"Women Ferry Pilots," p. 1; Knapp, *New Wings for Women*, pp. 55-56.

[15]Treadwell, *The Women's Army Corps*, p. 15 [quotes memo, Capt. Williston B. Palmer for G-1, 2 October 1939, sub: Women with the Army (Emergency), G-1/15839].

[16]Ibid., p. 16.

[17]Moolman, *Women Aloft*, pp. 141-143.

[18]Cochran and Odlum, *The Stars at Noon*, pp. 98-107.

[19]Ibid., p.107.

[20]Ibid., pp. 98-107; Keil, *Those Wonderful Women*, pp. 50-51, 97-100.

[21]Bradbrooke, "Atta Girls!" pp. 35, 44; Bowater, "Air Transport Auxiliary Service," pp. 172-173.

[22]Bradbrooke, "Atta Girls!" p. 73.

[23]Ibid., p. 44; Moolman, *Women Aloft*, p. 143.

[24]Treadwell, *The Women's Army Corps*, pp. 19-20, 23.

[25]Bandel, *The WAC Program*, p. 4 [quotes letter from General Headquarters, Air Force, to Chief of Air Corps, ACC 324.5 AWS (Women), dated 27 December 1941]. The Army Air Corps became the Army Air Forces on 20 June 1941; however, many individuals continued to refer to it as the Army Air Corps throughout the war.

[26]Bandel, *The WAC Program*, p. 15; see also Appendix III, Table 1.

[27]Craven and Cate, *The Army Air Forces*, p. 510; Bandel, *The WAC Program*, p. 9.

[28]Bandel, *The WAC Program*, p. 13 note.

[29]Craven and Cate, *The Army Air Forces*, p. 508.

[30]By contrast, the WAVES did not accept any black women during its first three years. (Treadwell, *The Women's Army Corps*, p. 58.) Neither did Marine Corps Women's Reserve (MCWR) nor the Coast Guard's women's corps (SPARs), although there was no explicit legislation preventing their doing so. (MacGregor, *Integration of the Armed Forces 1940-1965*, p. 74.)

[31]Treadwell, *The Women's Army Corps*, pp. 58-59, 596.

[32]The War Department restricted service by Japanese-American women until late in the war, when it wanted to recruit them (from relocation centers) to translate captured Japanese war documents. Virtually no one signed up, partly due to Japanese cultural opposition to working women, but primarily because of the Japanese-Americans' profound anger at the United States government for their wartime experience. (Treadwell, *The Women's Army Corps*, p. 589.)

[33]Hancock, *Lady in the Navy*, pp. 50-51.

[34]Ibid., pp. 51-52.

[35]Brecht, "Long May She WAVE," pp. 70, 78-79.

[36]Hancock, *Lady in the Navy*, pp. 53-56.

[37]Ibid., p. 61.

[38]Ibid., pp. 62-63.

[39]"WAVES," p. 15.

[40]Hancock, *Lady in the Navy*, pp. 271, 275-276.

[41]Holm, *Women in the Military*, pp. 64-65.

[42]Wirtschafter, interview.

[43]The one exception to this rule was Joy Bright Hancock. Because her job entailed extensive traveling between the various aviation schools of the Training Division in order to supervise the arrangements for the WAVES, she received official permission to fly Navy airplanes. One magazine article about her noted that "there are no soreheads among the men in the cockpits—for she talks their language." It is not clear, however, whether she was the pilot of these aircraft or simply a pilot who was permitted to occasionally sit at the controls. (Brecht, "Long May She WAVE," p. 79.)

[44]"WAVES," p. 13.

[45]"Rulers of the Air," p. 68.

[46]Hancock, "The Waves," p. 249.

[47]The MCWR program had been authorized 30 July 1942, and like the WAVES it had a World War I precedent, as women had served as Marine (F). (Holm, *Women in the Military*, pp. 33, 65; "Women Marines," p. 18.

[48]"Women Marines," p. 18.

[49]Holm, *Women in the Military*, pp. 32-33.

[50]Legislation permitting women to serve outside the continental United States was passed in September 1944; Hancock, *Lady in the Navy*, pp. 209-213, 269-270.

[51]Craven and Cate, *The Army Air Forces*, pp. 512, 528.

[52]Keil, *Those Wonderful Women*, pp. 101, 103.

[53]Crane, "The Women with Silver Wings," p. 8; La Farge, *The Eagle in the Egg*, pp. 60-61.

[54]Keil, *Those Wonderful Women*, p. 105.

[55]Ibid.; Margaret C. Love, interview.

[56]Cochran and Odlum, *The Stars at Noon*, p. 118.

[57]Ibid., pp. 117-118.

[58]"Mrs. Love of the WAFS," pp. 46, 51; Keil, *Those Wonderful Women*, p. 107; La Farge, *The Eagle in the Egg*, p. 131.

4. Nieces of Uncle Sam
(pages 44-56)

Epigraph: "The Wasp Songbook," compiled by members of Class 44-W-10, nd.

[1]Teague, "Memorandum on Miss Jacqueline Cochran."

[2]Ibid.

[3]Knowles, memorandum.

[4]Carter, "The Ladies Join the Air Forces," p. 96.

[5]Ibid., p. 88.

[6]Fort, "At the Twilight's Last Gleaming," p. 19.

[7]Crane, "The Women with Silver Wings," p. 10; Gillies, interview; Bohn, interview; Margaret C. Love, interview.

[8]Selby, "The Fifinellas," p. 76. There were three Soviet Air Force Regiments composed of women pilots. These women not only flew for their country's war effort but were actually engaged in combat. (Myles, *Night Witches*.) Information about these women was limited, so Cochran's statement, while not entirely accurate, did reflect the fact that the United States was the only country which established a major training program to teach women to fly military aircraft, although it was never possible for such a woman to "enlist" without some previous flight experience.

[9]Cochran, "Final Report."

[10]Ibid., pp. 6-10. Given the assumption that military flying was quite different from general aviation and the establishment's fear

that women might not be capable of mastering these unique skills, the reasons for the flight time requirement were two-fold. First, it was believed that a significant number of flying hours demonstrated a woman's genuine commitment to aviation, and further, it suggested that she possessed the requisite capabilities, such as mechanical aptitude and general knowledge of aeronautics. The second reason was Cochran's desire to have an outstanding "success to failure" ratio in her first classes. Using experienced pilots in the early phases of the program ensured a high level of achievement. Later, as the women proved themselves, the requirement was lowered.

[11]Pateman, "The WASP."

[12]Weisfeld, "The Role of the Women Airforce Service Pilots," p. 13; Cochran and Odlum, *The Stars at Noon*, p. 127. Black men were at least admitted into training programs and would eventually comprise four squadrons, the 99th, 100th, 301st, and the 302nd.

[13]Cochran and Odlum, *The Stars at Noon*, pp. 127-128; Hull, Scott, and Smith, *All the Women are White*, p. 21.

[14]Cochran and Odlum, *The Stars at Noon*, pp. 127-128.

[15]If candidates failed or left the program, they would have to pay for their trip home. (Strother, "The W.A.S.P. Training Program," pp. 299-301.)

[16]Strother, "The W.A.S.P. Training Program," pp. 299-301; Tanner, "We Also Served," p. 16.

[17]It should be noted that this does *not* apply to most WASPs.

[18]Felker, interview.

[19]Cochran, "Final Report," cover letter.

[20]Weisfeld, "The Role of the Women Airforce Service Pilots," pp. 20-21; Felker, interview.

[21]Cochran, "Final Report," p. 33.

[22]"The Reminiscences of Jacqueline Cochran" [quoted by permission from the Oral History Research Office, Columbia, University].

[23]Weisfeld, "The Role of the Women Airforce Service Pilots," pp. 24, 27 [quotes interview with S.R. Constance Howerton, 43-W-4].

[24]Ibid., pp. 28-29; Cochran and Odlum, *The Stars at Noon*, pp. 126-127.

[25]Weisfeld, "The Role of the Women Airforce Service Pilots," pp. 30-31; Pateman, interview.

[26]Cochran, "Final Report," p. 28. See Appendix III, Table 2.

[27]Cochran prevailed on this issue despite major protests from the Quartermaster General of the Army. Her work succeeded in overturning several regulations concerning uniforms for civilians as well as an advisory from the Air Judge Advocate that upheld the policy against issuing any uniform to nonmilitary personnel. An exception was authorized by the Comptroller General provided that the uniform was limited to outer garments and remained government property. (Risch, *A Wardrobe for the Women of the Army*, pp. 147-148; Cochran and Odlum, *Stars at Noon*, pp. 123-124.)

[28]"Unnecessary and Undesirable?" p. 66.

[29]Tunner, memorandum. See Appendix II.

[30]Keil, *Those Wonderful Women*, p. 268.

[31]Cochran and Odlum, *The Stars at Noon*, p. 121; "Battle of the Sexes," p. 71.

[32]Treadwell, *The Women's Army Corps*, p. 784.

[33]Knowles, memorandum; U.S. Congress, House, Select Subcommittee of the Committee on Veterans Affairs, *To Provide Recognition*, pp. 204-223.

[34]Cochran, "Final Report"; Gen. H.H. Arnold, memorandum.

[35]For costs per student see Appendix III, Table 3.

[36]Poole, "Requiem for the WASP," p. 55.

[37]By 1944, 6,500 nurses were assigned to the AAF, 6,000 at AAF station hospitals, 500 as flight nurses. Flight nurses served on aircraft in the evacuation of the wounded throughout the world. Nursing was the only established and accepted female branch in the military prior to World War II, but flight nursing was a new occupation. The nurses received special training and had to pass the flight surgeon's physical examination. (Craven and Cate, *The Army Air Forces*, p. 537.)

[38]Moolman, *Women Aloft*, p. 159.

[39]See Heilbrun, *Reinventing Womanhood* for a discussion of this concept.

5. Demobilization and the Postwar Transition
(pages 57-68)

Epigraph: *99s Newsletter* (15 April 1948):8.

[1]"Home by Christmas," pp. 68-69.

[2]"Women and Wrenches."

[3]Aircraft Industries Association of America, Inc. *Aviation Facts and Figures 1955*, pp. 17-27 [quotes U.S. Department of Commerce, Bureau of Labor Statistics, *Historical Statistics of the U.S.: Colonial Times to 1970*, Series D 29-41, p. 131]. Note that women's employment did decline after World War II from a peak of 19.3 million in 1945 to 16.8 million in 1946. However, the total male employment also decreased (3.2 million fewer men in 1946 than 1945). During this period the percentage of women in the labor force actually increased, which contradicts the idea that as a general rule women lost their jobs to men returning from the war.

[4]Aircraft Industries Association of America, Inc., *Aviation Facts and Figures 1953*, p. 61 [quotes Bureau of the Census, "Census of Manufactures, 1947, Aircraft and Parts," p. 3]; Bright, *The Jet Makers*, pp. 11-13.

[5]Aircraft Industries Association of America, Inc., *Aviation Facts and Figures 1953*, p. 61 [quotes Bureau of Labor Statistics, "Employment and Payrolls," (Monthly)].

[6]Gillies, interview.

[7]"To Serve on Vocational Advisory Committee," p. 64; Charles P. May, *Women in Aeronautics*, pp. 182-183.

[8]"Air-minded Miss," p. 23.

[9]Elizabeth Gardner, interview.

[10]"Ex-WASPs Ferrying Surplus War Training Planes," p. 1.

[11]Bohn, interview; Bohn, "Personal Data Record."

[12]Bright, *The Jet Makers*, pp. 77-79.

[13]"So You Want to Be a Hostess," p. 6.

[14]"Braniff Topnotchers," p. 10.

[15]Nielsen, *From Sky Girl to Flight Attendant: Women and the Making of a Union*, pp. 32-35, 37, 40. Nielsen's work is an excellent source for the story of the flight attendants.

[16]Ibid., pp. 41, 44-45, 48-49.

[17]Merryfield, "Five Hours to Solo," p. 20.

[18]Ibid.

[19]Dallimore, "Ceiling Unlimited," p. 2.

[20]Lempke, "President's Column," p. 1; The Ninety-Nines con-

ducted an informal survey during 1945 which indicated that nationally, flight time in a Cub trainer was averaging about $7 an hour for solo time and $10 for dual instruction. ("Ex-WASPs Ferrying Surplus War Training Planes," p. 1.)

[21]*99s Newsletter* (15 November 1945):1; (15 April 1946):1; (15 August 1946):1.

[22]Ibid. (15 March 1947):3; The Ninety-Nines, Inc., *The History of the Ninety-Nines, Inc.*, p. 32.

[23]*99s Newsletter* (15 April 1948):1.

[24]The Ninety-Nines, Inc., *The History of the Ninety-Nines, Inc.*, pp. 77-79.

[25]Noyes, "Again Women Fliers," p. 8. Betty Skelton, a champion aerobatic pilot who made her debut in the 1947 Air Show, had this advice for women pilots: "Be a lady." A contemporary account noted that Skelton "doesn't see any point to slopping about an airport in overalls" and that her standard flying luggage always included "an evening dress, a hat, and some nifty slack suits." (Fuller, "Betty Skelton Flies an Airshow," p. 76.)

[26]Downey, "Future Flyers," pp. 16-17.

[27]"Skymarker," p. 115; White, "The Sky's Their Limit," p. 328.

[28]Holm, *Women in the Military*, p. 103.

[29]"Wave Power in Aviation," p. 15.

[30]Treadwell, *The Women's Army Corps*, pp. 739-742, 747-748; Holm, *Women in the Military*, p. 114.

[31]Holm, *Women in the Military*, p. 113.

[32]Hancock, *Lady in the Navy*, p. 236.

[33]"Waves Have Good Record," p. 16; "Waves Join Reserves," p. 23; "Air Ambulences Fly Men to Hospitals," p. 20.

[34]MacGregor, *Integration of the Armed Forces*, pp. 248, 267 [p. 267 quotes letter, A. Philip Randolph to General C.B. Cates, 8 March 1949, and letter, CMC to Randolph, 10 March 1949, AW828].

[35]Holm, *Women in the Military*, pp. 130, 132.

[36]Rasmussen, interview.

[37]Jesse J. Johnson, *Black Women in the Armed Forces*, p. 23.

6. "The Feminine Mystique" and Aviation
(pages 69-80)

Epigraph: American Airlines publication, ca. 1950.

[1]Crist, "Operation Polar," p. 8.

[2]Ibid.

[3]Kraft, "Flying in the Face of Age," p. 30; Vetterlein, "Newfoundland to Ireland, Non-Stop," p. 18; Buck, "The Most Unforgettable Character," p. 115; Hart, "She Flew the Atlantic," p. 76.

[4]Buck, "The Most Unforgettable Character," p. 117.

[5]"Miss Cochran Holds Most Jet Records," p. 26; "Cochran Sets Sights," p. 18. Women were not allowed to fly USAF aircraft. Cochran was no exception to this rule. She arranged to be hired by aircraft companies that manufactured military aircraft. In the case of this record, Cochran flew an airplane owned by the Canadian government. The USAF did not place any restrictions on the pilots employed by companies or other governments, which explains how it was possible for Cochran to make this flight.

[6]Cochran and Odlum, *The Stars at Noon*, pp. 221-243.

[7]Holm, *Women in the Military*, pp. 141-144.

[8]Geraldine May, letter.

[9]Holm, *Women in the Military*, pp. 148-157.

[10]"Air Force Executive," p. 13.

[11]Wolfe, "Women and the Nation's Security," p. 13; Thruelsen, "Flying WAF," pp. 28-29, 137.

[12]"The Glamour Corps," pp. 99, 149.

[13]"Woman Fills Many Roles at Sheppard," p. 19.

[14]"Women in Aviation," p. 6.

[15]"Gals Try Their Hands," p. 33.

[16]"The Lady Is Also a Wave," p. 24.

[17]Holm, *Women in the Military*, pp. 152-153.

[18]"Navy Flight Nurses Care for Wounded," pp. 19-20.

[19]"DACOWITS History Update," pp. 1-22.

[20]"A Record of DACOWITS," pp. 1-2.

[21]Phillips, "Readin', Writin', and RPM's," pp. 21, 48.

[22]"Top-flight Scholar," p. 55.

[23]Burnham, "U.S. Ladies in the Air," pp. 32, 37. There was an assumption that women might not take the CAP as seriously as men. Note comments such as Burnham's (p. 33): "Even when doing a man's job, Harriet enjoys her woman's prerogative—first to get a steaming cup of hot coffee."

[24]Burnham, "The Defense Department's First Lady of Flight," p. 15.

[25]"Lady Flier Completes 10-day Tour," p. 29; Nichols, *Wings for Life*, pp. 250-282, 309.

[26]Brick, *Powder Puff Derby*, p. 36.

[27]Ibid., p. 6.

[28]Ibid., pp. 6-7; Brick, "Million Dollar Race," pp. 46-47, 74, 76.

[29]"TAR," p. 44. There was a contest rule after 1952 requiring all contestants to wear dresses or suits. Slacks or shorts were not permitted.

"This regulation stems from the same desire for serious recognition that has led the TAR participants to shun the title 'Powder Puff Derby,' which was originally applied to their big race." (Wolfe, "Women's Air Race," p. 50.)

[30]Brick, *Powder Puff Derby*, p. 20.

[31]Ibid., p. 28.

[32]In 1959, *Newsweek* described the AWTAR as follows: "This is the way of women—or at least 100 American women—in the aviation age. Once each year, they shed their household duties, climb into tiny aircraft and with unladylike abandon, race from one coast of the U.S. to the other. They zoom in and out of obscure airports, scream at attendants to fill up their gas tanks, and roar through treacherous rainstorms. Since this annual female aerial madness began in 1947, it has become known as the Powder Puff Derby." The description was erroneous and overblown but it also betrayed a certain lack of seriousness on the part of the editorial staff, and revealed something of the underlying attitude toward women in aviation. ("Powderpuff Derby," p. 88.)

[33]Eddleman, *Cows on the Runway*, p. 172.

[34]Ingells, "Their Eggbeaters Aren't in the Kitchen," p. 7.

[35]Dougherty, letter; Charles P. May, *Women in Aeronautics*, p. 208.

[36]Aerospace Industries Association of America, Inc., *Aviation Facts and Figures 1959*, pp. 70, 73, 75.

[37]"Unlimited Opportunities for Women," p. 9.

[38]"Designing Women," p. 13. Out of 2,100 engineering students at the University of Washington in Seattle (right next to Boeing)

only 10 were women.

[39]Ibid.

[40]Aerospace Industries Association of America, Inc., *Aerospace Facts and Figures 1961*, p. 83 [quotes Department of Labor, Bureau of Labor Statistics, "Employment and Earnings"].

[41]Aerospace Industries Association of America, Inc., *Aviation Facts and Figures 1959*, p. 77; U.S. Department of Commerce, *Historical Statistics of the United States: Colonial Times to 1970, Part I*, p. 298.

[42]Flynn, "Ladies with the Last Word," p. 17. Also see Ida F. Davis, "The Lady Finally Hacked It!" pp. 16-17.

[43]"Topside Aviation Club," p. 32.

[44]"First All-Women's A & E Mechanics Course," p. 1.

[45]Chase, *Skirts Aloft*, pp. 79-81.

[46]Nielsen, *From Sky Girl to Flight Attendant: Women and the Making of a Union*, p. 61; Nielsen, "From 'Sky Girl' to Flight Attendant: A Proud Union Legacy," p. 6.

[47]Nielsen, "From 'Sky Girl' to Flight Attendant: A Proud Union Legacy," p. 6.

[48]Wilson, "Salute to the Ninety-Nines," p. 22.

7. The Impact of the Women's Rights Movement
(pages 81-92)

Epigraph: Title VII, PL 88-352 (Civil Rights Act of 1964).

[1]"The Reminiscences of Ruth Nichols," p. 43.

[2]Cobb and Rieker, *Woman into Space*, pp. 129-135.

[3]"Follow Up on the News," p.33; "The U.S. Team," pp. 32-33.

[4]Cochran, "Women in the Space Age."

[5]Cochran, letter.

[6]Dryden, letter, 26 April 1962; Dryden, letter, 30 July 1962.

[7]Webb, letter; Lyndon B. Johnson, letter; Werne, "For and About Women," pp. 9-10; Cochran, letter.

[8]Luce, "But Some People," p. 31.

[9]The Civil Rights Act also made discrimination on the basis of race, color, religion, and national origin illegal. For further discussion see Beatty and Beatty, "Job Evaluation and Discrimination," pp.212-213.

[10]Friedan, *The Feminine Mystique*, p. 367.

[11]Enloe, *Does Khaki Become You*, pp. 188-190.

[12]Fitzroy, *Career Guidance for Women*, pp. 16-17, 124, 142. This percentage represents a change, albeit small, from the beginning of the time period covered by the present study. In 1938 there were less than a dozen women in all fields of engineering, according to statistics compiled by Margaret Rossiter. (Rossiter, *Women Scientists in America*, p. 136., 1982).

[13]Fitzroy, interview.

[14]Parrish, *Women in Engineering*, p. 5.

[15]Dietrich, "The Pilot is a Lady," p. 50.

[16]Ibid., p. 105.

[17]Horowitz, "Aviation Careers," p. 51.

[18]Demarest, "Just a Little Squeeze, Please," p. 44.

[19]Parke, "The Feminine Case," p. 28.

[20]Horowitz, "For Men Only?" pp. 30-33.

[21]Nielsen, *From Sky Girl to Flight Attendant: Women and the Making of a Union*, p. 81 [quotes Kelly Rueck, "A Time of Change," *Flightlog*, June 1973, pp. 2, 3, 6].

[22]Saunders, *So You Want to be an Airline Stewardess*, pp. 11, 164; Nielsen, *From Sky Girl to Flight Attendant: Women and the Making of a Union*, p. 83.

[23]Nielsen, *From Sky Girl to Flight Attendant: Women and the Making of a Union*, pp. 85-86.

[24]Friedan, *The Feminine Mystique*, p. 372.

[25]Nielsen, *From Sky Girl to Flight Attendant: Women and the Making of a Union*, pp. 89-90.

[26]Ibid., p. 90.

[27]U.S. Department of Transportation, FAA, U.S. Civil Airman Statistics 1960-1969.

[28]Merriam [Smith], "I Flew Around the World Alone," p. 77. [This article was signed "Joan Merriam," but all her awards and honors refer to her as "Joan Merriam Smith."]

[29]James Gilbert, "The Loser," pp. 80-84; "Shades of Amelia," pp. 20-21.

[30]Pellegreno, "I Completed Amelia Earhart's Flight," p. 48; The Ninety-Nines, Inc., *The History of the Ninety-Nines, Inc.*, p. 40.

[31]Buegeleisen, "Confessions of a Powderpuffer," p. 62.

[32]Tully, "Those Whirly-Girls," p. [1].

[33]U.S. Department of Transportation, FAA, *History of the Women's Advisory Committee on Aviation*, 1967, pp. 1-2.

[34]Buegeleisen, "Skirts Flying," p. 32; "Traffic Pattern," pp. 19-20; Howard, interview.

[35]Holm, *Women in the Military*, p. 183.

[36]Ibid., pp. 177, 184.

[37]Ibid., p. 202.

[38]"VF-126's Historical First," p. 2.

[39]"Wave Solos in T-34 Mentor," p. 33.

[40]Morris, "Service Women Get Shot," p. 34; "Another Wave 'First' Cited," p. 37.

[41]Jesse J. Johnson, *Black Women in the Armed Forces*, pp. 23, 33.

[42]Holm, *Women in the Military*, pp. 205, 210, 223, 227; Pateman, interview.

[43]Callander, "Why Can't a Woman Be a Military Pilot?" p. 13.

[44]Ibid.

[45]Buegeleisen, "Skirts Flying," p. 24.

8. Women with "The Right Stuff"
(pages 93-106)

Epigraph: U.S. Congress, House, Select Subcommittee of the Committee on Veterans' Affairs, *To Provide Recognition*, p. 277.

[1]Aerospace Industries Association of America, Inc., "Employment of Women in Aerospace," memoranda from 22 October 1979 and 14 April 1980.

[2]The term "light" refers to the mass of the end product. Manufactured goods are usually described as being either "light" or "heavy," "durable" or "nondurable." Sometimes the industry that provides such a product is also labled "light" or "heavy." Light industries such as textiles, garments, food processing, or electronics traditionally have had higher concentrations of women in their labor force. The perceived inability of women to meet certain physical standards in conjunction with the stereotypes about women's ability to excel (such as dexterity or the ability to endure tedium and repetition) work to channel women into these particular areas of employment.

[3]Bruer, "Women in Science," p. 3 [quotes National Research

Council, *Climbing the Academic Ladder II* (Washington, D.C., National Academy of Sciences, 1983)]; Fitzroy, *Career Guidance,* p. 16.

[4]Ruina, *Women in Science and Technology,* pp. 13, 21.

[5]Howard, "A Salute to Women in Aerospace," pp. [2-3].

[6]"First Lady of Aviation," p. 10.

[7]Cardozo, "Athley Gamber," p. 22.

[8]Elizabeth Simpson Smith, *Breakthrough,* p. 150.

[9]"Air-Traffic Controllers," p. 64. From 1977 to 1980 the total number of controllers remained nearly constant, while the presence of women rose from 4.2% to 5.3% (U.S. Department of Transportation, *FAA Statistical Handbook.*)

[10]Howard, "A Salute to Women in Aerospace," p. [4]. In the Department of Transportation there were 14,864 engineers in 1973; 72 (or 0.5%) were women. For comparison, both the military and NASA employed similar numbers of women. NASA had 110 (0.8%) of 14,193; the Air Force 144 (0.9%) of 15,245; the Navy 386 (1%) of 39,363; the Army, 475 (1.4%) of 33,025. All other women engineers were outside of aerospace. (Fitzroy, *Career Guidance,* p. 131.)

[11]For 10 months in 1934 and 1935, Helen Richey worked as a copilot for Central Airlines. (Oakes, *United States Women in Aviation: 1930-1939,* p. 13.) Also see: "Women in the Airline Cockpit," p. 95; Elizabeth Simpson Smith, *Breakthrough,* p. 150.

[12]Schweider, "Emily Howell," p. 55.

[13]Kanner, "Women Airline Pilots," pp. 33, 35.

[14]Schweider, "Emily Howell," p. 55.

[15]Kanner, "Women Airline Pilots," p. 37.

[16]Ibid.

[17]Burgen, "Winging It at 25,000 Feet," pp. 58, 60.

[18]U.S. Department of Transportation, FAA, Office of Aviation Policy, *Women in Aviation and Space,* p. 13.

[19]U.S. Department of Transportation, FAA, Office of Aviation Policy, *Flight Attendants,* p. 5.

[20]Nielsen, *From Sky Girl to Flight Attendant: Women and the Making of a Union,* pp. 113-114, 131-132.

[21]Ibid., p. 115.

[22]Ibid., pp. 120-121; Nielsen, "From 'Sky Girl' to Flight Attendant: A Proud Union Legacy," p. 6.

[23]Holm, *Women in the Military,* pp. 246-250; "Skirts for Twenty Percent," p. 16.

[24]Holm, *Women in the Military,* pp. 250, 313.

[25]Collins, "From Plane Captains to Pilots," pp. 9, 12.

[26]"Young, Successful, and First," p. 53.

[27]Shipman, "The Female Naval Aviator," p. 84.

[28]Collins, "From Plane Captains to Pilots," pp. 13-14.

[29]Ibid., p. 16; Farrell, "... and Navy Women Today," p. 28.

[30]Collins, "From Plane Captains to Pilots," pp. 17-18.

[31]"Army's 1st Female Pilot," p. 43.

[32]Streeter and Hamilton, "Womanpower in Dungarees," p. 22.

[33]Ibid.

[34]Collins, "From Plane Captains to Pilots," p. 15. Not until 1979 were there any black women pilots in the military. Second Lt. Marcella A. Hayes graduated then from the Army Aviation School. ("Air Force Graduates," p. 46.)

[35]"Distaff Pilots," p. 3; "Women Pilots," p. 47.

[36]"Test Pilot School," p. 26.

[37]"Women Soon to Train," p. 2.

[38]Famiglietti, "Female Flier Plans Shaping Up," p. 4.

[39]"Six Women to Enter Navy Training," p. 22.

[40]Famiglietti, "Planes Listed for Women Pilots," p. 11.

[41]Familglietti, "Women Named to Undergo UPT," p. 30.

[42]Terry Arnold, "Baptizing the New Breed," p. 5.

[43]Farrell, "... and Navy Women Today," p. 28; "Young, Successful, and First," pp. 53, 134.

[44]Terry Arnold, "Baptizing the New Breed," pp. 6-7.

[45]Eaker, "Combat Duty for Women," p. 17; Jenkins, "WAVES Established Women," p. 1.

[46]U.S. Department of Transportation, FAA, Office of Aviation Policy, *Women in Aviation and Space,* p. 8.

[47]Stiehm, *Bring Me Men and Women,* pp. 18-20.

[48]Holm, *Women in the Military,* pp. 309-310.

[49]"Brief History on WASP Struggle," pp. [3-4].

[50]Keil, *Those Wonderful Women,* pp. 311, 315-316.

[51]Ibid., p. 313.

[52]U.S. Department of Transportation, FAA, *Civil Airmen Statistics 1970-1980.* See also Appendix III, Table 4.

[53]Novello and Youssef, "Psycho-Social Studies in General Aviation: II," pp. 631, 633.

[54]Phinzy, "Mary, Mary Quite Contrary," p. 80. Gaffaney was later awarded the Lady Hay Drummond-Hay–Jessie R. Chamberlin Memorial Trophy by the Women's International Association of Aeronautics. (The Ninety-Nines, Inc., *The History of the Ninety-Nines, Inc.,* p. 45.)

[55]Howard, "The Whirly-Girls—Flying Ambassadors," p. 22.

[56]The Ninety-Nines, Inc., *The History of the Ninety-Nines,* p. 45.

[57]Gasnell, "The Last Powder Puff," p. 8.

[58]Zirker, "Moline, Illinois, July 9," pp. 105-106.

[59]The Ninety-Nines encountered a problem in trying to serve the woman pilot who was working full time. Unable to attend meetings in the middle of the day, these women felt different from the "flying housewife" and a small split in the organization began to develop. Although flying still bonded these women together, their life experiences were quite different. The Ninety-Nines' solution, in keeping with its grass-roots orientation, has been simply to assist in the formation of new chapters. Thus in metropolitan regions it is not uncommon to find two or three chapters, each with its own distinct personality, each addressing the needs of its membership.

9. Captains of Industry, Airlines, and the Military
(pages 107-114)

Epigraph: Christman, "Navy's First Female Test Pilot," p. 26.

[1]Aerospace Industries Association of America, Inc., "Employment of Women in Aerospace," 6 August 1985 update.

[2]Aerospace Industries Association of America, Inc., *Aerospace Facts and Figures 1985/86,* p. 154.

[3]"It's Women's Week," p. 1.

[4]Milton, "Gayle Ranney," pp. 96, 99, 100.

[5]Meyers, letter, 11 June 1985; Meyers, letter, 23 July 1985.

[6]"Meet Your Instructors," p. 2.

[7]Washburn, "Professional Women Controllers," p. 14.

[8]"Sky High About Her Job," p. 1.

[9]There is a social organization of women airline pilots known as the International Social Affiliation of Women Airline Pilots (the

group's nickname is ISA + 21). Its function is to provide a network of communication among the women and a high degree of support.

[10]Rippelmeyer, interview.

[11]Stella Smith, "Lynn Rippelmeyer," p. 17.

[12]"Coffee, Tea or Customer Service Managers?" p. D5.

[13]Nielsen, *From Sky Girl to Flight Attendant: Women and the Making of a Union*, p. 136.

[14]Conant et al., "Women in Combat?" p. 36.

[15]Burdette, "Making it in the Military," p. 18.

[16]In Grenada some of the women were in the actual combat area. Also, Army women fly in the Demilitarized Zone in Korea and other high combat probability areas. The Army does not have the same combat restrictions as the Air Force and the Navy.

[17]Conant et al., "Women in Combat?" p. 38.

[18]Mohler, "Sport Pilot Medicine," p. 61.

[19]Nelson, "Women in Hang Gliding," p. 16; de Man, "She's Magnificent in Her Flying Machine," p. 36.

Glossary of Abbreviations

AAF	Army Air Forces
AAFTD	Army Air Forces Flying Training Detachment
AF	Air Force (United States)
AFA	Association of Flight Attendants
AGARD	Advisory Group for Aerospace Research and Development, part of the North Atlantic Treaty Organization (NATO)
Air WAC	Member of the Women's Army Corp (WAC) serving with the Army Air Forces (AAF)
ALPA	Air Line Pilot's Association
ALSA	Air Line Stewardess Association
ALSSA	Air Line Stewards and Stewardesses Association
ATA	Air Transport Auxiliary (Great Britain)
ATC	Air Transport Command
ATO	Air Transport Officers
AWS	Aircraft Warning Service
AWTAR	All-Woman Transcontinental Air Race
AWVS	American Women's Voluntary Services, Inc.
BuAer	Bureau of Aeronautics, United States Navy
CAA	Civil Aeronautics Administration (Civil Aeronautics Authority prior to 1940)
CAP	Civil Air Patrol
CPTP	Civilian Pilot Training Program
CO	Commanding Officer
DACOWITS	Defense Advisory Committee on Women in the Services
FAI	Fédération Aéronautique Internationale
IUFA	Independent Union of Flight Attendants
MATS	Military Air Transport Service
MCWR	Marine Corps Women's Reserve (United States)
MOS	Military Occupational Specialty
NAA	National Aeronautic Association
NASA	National Aeronautics and Space Administration
NASM	National Air and Space Museum, Smithsonian Institution
NATO	North Atlantic Treaty Organization
NATS	Naval Air Transport Service
NIFA	National Intercollegiate Flying Association
99s	Ninety-Nines, Inc. International Women Pilots Association
NOW	National Organization for Women
NWLB	National War Labor Board
PWC	Professional Women Controllers, Inc.
RAF	Royal Air Force
ROTC	Reserve Officer Training Program
SPARs	Semper Paratus—Always Ready (U.S. Coast Guard motto and the name given to the women's corps of the Coast Guard during World War II)
SWE	Society of Women Engineers
TWU	Transport Workers Union
USA	United States Army
USAF	United States Air Force
USN	United States Navy
VA	Veterans Administration
WAAC	Women's Army Auxiliary Corps
WAC	Women's Army Corps
WACOA	Women's Advisory Committee on Aviation
WAF	Women in the Air Force
WAFS	Women's Auxiliary Ferry Squadron
WAMS	Women Apprentice Mechanics
WASP	Women's Airforce Service Pilots
WAVES	Women Accepted for Voluntary Emergency Service (Initially it was "Woman Appointed ..."until it was recognized that only officers were appointed, and the term was changed to "accepted."
WFA	Women Flyers of America, Inc.
WFTD	Women's Flying Training Detachment
WNAA	Women's National Aeronautical Association
WTS	War Training Service

References

This study is an introductory survey of the topic. It highlights events and individuals, as appropriate, but its main purpose is to examine the critical questions and issues in the history of women in aviation from 1940 to 1985. There is a limited secondary literature, which tends to draw on the same small number of well-known primary documents. The prospective researcher in this area need not be discouraged, however; the primary sources for the study of women in aviation are both abundant and available.

Unless credited otherwise, all interviews were conducted by the author (D.G.D.). Copies of all taped interviews are held by the NASM Archival Support Center at the Paul Garber Facility, Suitland, Maryland.

Aeronautical Chamber of Commerce of America. *Aircraft Yearbook, 1946.* New York: Lanciar Publications, 1946.

Aerospace Industries Association of America, Inc. *Aviation Facts and Figures 1959.* Washington: American Aviation Publications, 1959. This is a continuation of the same title published prior to 1959 by Aircraft Industries Association of America, Inc. Subsequent volumes for 1961 through 1985/1986 were issued as *Aerospace Facts and Figures,* various publishers.

——. "Employment of Women in Aerospace." Statistical Series from 22 October 1979 to present. On file in A.I.A.'s Economic Data Service Library, Washington, D.C.

Ahnstrom, Doris N. "Look ... Women." *Skyways,* 2 (September 1943):30-32, 66.

"Air Ambulences Fly Men to Hospitals." *Naval Aviation News* (August 1946):20-21.

"Air Force Executive." *National Business Woman,* 36 (June 1957):12-14.

"Air Force Graduates First Black Woman Pilot." *Ebony* (January 1983):46.

"Air-minded Miss." *Flying,* 40 (May 1947):22-23.

"Air-Traffic Controllers." *Working Woman,* 4 (February 1979):63-64.

Aircraft Industries Association of America, Inc. *Aviation Facts and Figures 1945.* New York: McGraw-Hill Book Co., Inc., 1945. There were issues of this title for 1953, 1955-1958, various publishers. This title did not become an annual publication until 1955. After 1958, it was compiled by Aerospace Industries Association of America, Inc.

"Alaska—Thumbs Up." *New Horizons,* XIII (April 1943):19.

"The Alleged Weaker Sex." *Independent Woman,* 19 (October 1940):316.

"America's Interesting People: 4-M, Arlene Davis." *American Magazine,* 129 (February 1940):67.

Anderson, Karen. *Wartime Women: Sex Roles, Family Relations, and the Status of Women.* Westport, Connecticut: Greenwood Press, 1981.

"Ann Devers Joins NBAA Staff as Manager, Educational Programs." National Business Aircraft Association press release, 22 September 1985.

"Another Wave 'First' Cited." *Naval Aviation News* (July 1967):37.

"Army's 1st Female Pilot Wins Helicopter Wings." *Army,* 24 (July 1974):43.

Arnold, General H.H. Memorandum to General Giles on WASP after demobilization, 5 October 1944. Nancy H. Love Papers, NASM Archives, Washington, D.C.

Arnold, Terry. "Baptizing the New Breed." *Airman,* 21 (October 1977):2-7.

"Around the World in 33 Days." *Air Pictorial,* 28 (August 1966):297.

Arthur, Julietta K. "Airways to Earning." *Independent Woman,* 19 (February 1940):34-35, 55-56.

——. "Now You Can Learn to Fly." *Independent Woman,* 19 (October 1940):320-321, 336.

——. "Wings for the Working Girl." *Flying,* 31 (December 1942): 41-42, 136-137.

Ash, Mae, and Jerome Beatty. "I Let My Daughter Fly." *American Magazine,* 130 (September 1940):32-33, 131-133.

"Babies, Just Babies." *Flying Time* (9 March 1940).

Bach, Richard. "The Invisible 99s." *Flying,* 77 (August 1965):39-41.

——. "Richard Bach: An Airplane for Bette." *Flying,* 84 (April 1969):32-33.

Backus, Jean L. *Letters From Amelia: An Intimate Portrait of Amelia Earhart.* Boston: Beacon Press, 1982.

Bacon, Marion. "Aviatrix in Blue." *Airman,* 16 (January 1972): 31-32.

Bailey, Mildred C. "Army Women and a Decade of Progess." *Army,* 24 (October 1974):85-91.

Baker, Laura Nelson. *Wanted: Women in War Industry: The Complete Guide to a War Factory Job.* Introduction by Elinore M. Herrick. New York: E.P. Dutton & Co., Inc., 1943.

Bandel, Betty. *The WAC Program in the Army Air Forces.* Washington: Military Personnel Division, Assistant Chief of Air Staff-1, Army Air Forces Headquarters, 1945.

Barger, Harold. *The Transportation Industries 1889-1946: A Study of Output, Employment, and Productivity.* New York: National Bureau of Economic Research, Inc., 1951.

Barnard, Richard C. "Jackie Cochran: Famed Pilot Says Women

Shouldn't Fight." *The Times Magazine* (supplement to *Air Force Times*), 38 (23 January 1978):18, 20.

Bates, Stephen. "Lady with Wings." *Popular Aviation* (May 1940):55-56, 88.

"Battle of the Sexes." *Time*, 43 (8 May 1944):68, 71.

Baxter, Gordon. "The Bax Seat: The T-Shirt 206." *Flying*, 103 (December 1978):112.

———. "Cochran." *Flying*, 101 (September 1977):178.

Beatty, Richard W., and James R. Beatty. "Job Evaluation and Discrimination: Legal, Economic, and Measurement Perspectives on Comparable Worth and Women's Pay." *In* H. John Bernardin, editor, *Women in the Work Force*, pp. 205-234. New York: Praeger, 1982.

Bell, Susan E. "PMS and the Medicalization of Menopause: A Sociological Perspective." *In* B. Ginsburg and B.F. Carter, editors, *The Premenstrual Syndrome: Legal and Ethical Implications*, pp. 151-172. New York: Plenum Press, 1987.

Bender, Rose Marie "Bonnie" C. Telephone interview, November 1985.

Bernardin, H. John, editor. *Women in the Work Force*. New York: Praeger, 1982.

Binkin, Martin. *America's Volunteer Military: Progress and Prospects*. (Studies in Defense Policy.) Washington: The Brookings Institution, 1984.

———, and Shirley J. Bach. *Women in the Military*. (Studies in Defense Policy.) Washington: The Brookings Institution, 1977.

Black, Thomas H. "Pros' Nest, I've Got It." *Flying*, 102 (May 1978):22.

Blackmore, Orra Heald. "This is How It Began." *Skylady*, 2 (March-April 1946):27.

"Blonde First." *New Horizons*, XIII (November 1942):15.

"Board to Tap Women in July." *Air Force Times*, 36 (19 January 1976):10.

Boase, Wendy. *The Sky's the Limit: Women Pioneers in Aviation*. New York: MacMillian Publishing Co., Inc., 1979.

Bohn, Delphine [WAFS]. Interview, Washington, D.C., 23 January 1986.

———. "Personal Data Record." Autobiographical memorandum sent to D.G.D., 5 May 1986.

Bowater, Eric V. "Air Transport Auxiliary Service." *Flying and Popular Aviation*, 31 (September 1942):172-173, 260.

Bowman, Constance. *Slacks and Callouses*. New York: Longmans, Green and Co., 1944.

Boyne, Walter, J. "Blame It on Women." *AOPA Pilot* (October 1984):98-100.

Bradbrooke, Joan. "Atta Girls! American Girls Join the ATA to Ferry Britain's Fighters." *Skyways*, 2 (January 1943):34-35, 44-45, 73.

Bradley, La Verne. "Women in Uniform." In *Insignia and Decorations of the U.S. Armed Forces*, revised edition, pp. 159-160, 185-198. Washington: National Geographic Society, 1944.

"Braniff Topnotchers." *Skylady*, 2 (May-June 1946):10-11.

Brecht, Raymond C. "Long May She WAVE." *Skyways*, 2 (April 1943):19, 70, 78-79.

Brick, Kay. "Million Dollar Race." *Flying*, 64 (June 1959):46-47, 74, 76.

———, editor. *Powder Puff Derby: The Record 1947-1977*. Fallbrook, California: AWTAR, Inc., 1985.

"Brief History on WASP Struggle for Militarization!" *WASP Newsletter*, XIII (December 1976):[3-4].

Bright, Charles D. *The Jet Makers: The Aerospace Industry from 1945 to 1972*. Lawrence, Kansas: The Regents Press of Kansas, 1978.

Brooks-Pazmany, Kathleen L. *United States Women in Aviation, 1919-1929*. (Smithsonian Studies in Air and Space, 5.) Washington: Smithsonian Institution Press, 1983.

Brown, Betty. "Women Fliers are Active in War Effort." *Cleveland Plain Dealer* (19 July 1943).

Brown, Gwilym S. "With a Huff and a Puff." *Sports Illustrated*, 39 (30 July 1973):50.

Brown, Laura. "Flier or No, Girl Can Fill a Vital War Role." *New York World-Telegram* (23 January 1943):8.

Bruce, Jeannette. "Fly Away on Ladies' Day." *Sports Illustrated*, 35 (19 July 1971):44, 47.

Bruer, John T. "Women in Science: Toward Equitable Participation." *Science, Technology and Human Values*, 9 (Summer 1984):3-7.

Buck, Robert N. "The Most Unforgettable Character I've Met." *Reader's Digest*, 70 (August 1957):114-118.

Buegeleisen, Sally. "Abbott's Rib." *Flying*, 77 (August 1965):35-37, 98, 100, 102, 104, 106, 108.

———. "Confessions of a Powderpuffer." *Flying*, 81 (December 1967):58-63.

———. "Skirts Flying." *Flying*, 78 (February 1966):24. Also other columns of the same title in subsequent issues of *Flying*: volume 78 (March 1966):32, (April 1966):30, (May 1966):24-25; volume 79 (July 1966):28, (September 1966):22, 138; volume 80 (January 1967):74, (March 1967):101, (April 1967):16, (June 1967):28; volume 81 (October 1967):22, (November 1967):26, (December 1967):24.

Burdette, Liz. "Making It in the Military." *99 News*, 11 (June 1984):17-19.

Burgen, Michele. "Winging It at 25,000 Feet." *Ebony*, 33 (August 1978):58-60, 62.

Burnham, Frank. "The Defense Department's First Lady of Flight." *U.S. Lady*, 2 (Spring 1958):14-15.

———. "U.S. Ladies in the Air." *U.S. Lady*, 1 (April 1956):32-37.

Callander, Bruce. "Why Can't a Woman Be a Military Pilot?" *Air Force Times*, 27 (9 August 1967):13.

———. "Women in the Military: An Ongoing Revolution." *Air Force Times* (20 December 1982):8.

Cardozo, Yvette. "Athley Gamber: Flying High in Aircraft Sales and Service." *Working Woman*, 5 (January 1980):20-22.

Carter, Rowland. "The Ladies Join the Air Forces." *Flying*, 31 (December 1942):67, 88, 96.

Chapelle, Georgette L. [pseud. Dickey Meyer] *Girls at Work in Aviation*. New York: Doubleday, Doran and Co., Inc., 1943.

———. *Needed—Women in Aviation*. New York: Robert M. McBride & Co., 1942.

Chase, Lucille. *Skirts Aloft*. Chicago: Louis Mariano, 1959.

Chirchman, Deborah. "Flying Them or Fixing Them She Loves Planes." *Chronicle-Telegram* (9 September 1984).

Christman, Timothy J. "Navy's First Female Test Pilot." *Naval Aviation News* (November-December 1985):24-26.

Chun, Victor K. "The Origin of the WASPs." *American Aviation Historical Society Journal*, 14 (Winter 1969):259-262.

"The Civilian Pilot Training Program." *Civil Aeronautics Journal*,

1(1 January 1940):3-4.

Cobb, Jerrie, and Jane Rieker. *Woman into Space: The Jerrie Cobb Story.* Englewood Cliffs, New Jersey: Prentice-Hall, Inc., 1963.

"Cochran Sets Sights on 700 mph Record." *Aviation Week,* 58 (18 May 1953):18.

Cochran, Jacqueline. "Final Report on Women Pilot Program." Memorandum to Commanding General, Army Air Forces [1945]. Women Air Force Service Pilots, organization file, NASM Library, Washington, D.C.

———. Letter to Jerrie Cobb, 23 March 1962. Jerrie Cobb file, NASA History Office, Washington, D.C.

———. "A Woman's Views on Aviation." *Consolidator* (December 1940):29, 98.

———. "Women in the Space Age." Speech made to Zonta Club of Cleveland, Ohio, 28 November 1962. Copy in Jackie Cochran file, NASA History Office, Washington, D.C.

———, and Floyd Odlum. *The Stars at Noon.* Boston: Little, Brown and Co., 1954.

"Coffee, Tea or Customer Service Managers?" *Washington Post* (31 January 1986):D5.

Collins, Helen F. "Fifinella and Friends." *Naval Aviation News* (July 1977):21-23.

———. "From Plane Captains to Pilots: Women in Naval Aviation." *Naval Aviation News* (July 1977):8-18.

Conant, Jennet, et al. "Women in Combat?" *Newsweek* (11 November 1985):36-38.

Corn, Joseph J. *The Winged Gospel: America's Romance with Aviation, 1900-1950.* New York: Oxford University Press, 1983.

"Coup for Cochran." *Newsweek,* 22 (19 July 1943):40-41.

Covey, Ralph F. "'Equal Rights, Equal Risks.'" *U.S. Naval Institute Proceedings,* 112 (January 1986):16, 19.

"Crafts Shop Fills Idle Hours." *Naval Aviation News* (February 1946):23.

Crandell, Susan. "Karen Coyle: Speaking on Our Behalf." *Flying,* 104 (April 1979):97.

Crane, Mardo. "The Women with Silver Wings." *99 News,* 5 (Special Issue, 1978):8-13.

Craven, Wesley Frank, and James Lea Cate, editors. *The Army Air Forces in World War II.* (Services Around the World, 7). Chicago: University of Chicago Press, 1958. [See especially Sction V, "Women in the AAF," by Kathleen Williams Boom.]

Crist, Judith. "Operation Polar." *Senior Scholastic,* 63 (16 September 1953):8.

Crook, Bette. "Pilot? Who, Me?" *Flying,* 62 (February 1958):40-41, 88-89.

Crouch, Tom D. *The Eagle Aloft: Two Centuries of the Balloon in America.* Washington: Smithsonian Institution Press, 1983.

Cubbedge, Robert. "Vicki Frankovich: Toe-to-Toe with Carl Icahn." *Airport Press,* 7 (November 1985):19.

"DACOWITS History Update." Memorandum, 1981. DACOWITS.

Dallimore, Gladys. "Ceiling Unlimited ...." *Skylady,* 1 (March-April 1945):2.

"Dangers to Female Pilots to Be Checked on Planes." *Air Force Times,* 37 (21 March 1977):2.

Davis, Dorothy H. "A Wasp's Eye-View of Pioneer Days at Chico State University." Unpublished article, received by D.G.D., 4 May 1985.

Davis, Ida F. "The Lady Finally Hacked It!" *Flying,* 61 (August 1957):36-37, 90-91.

"D.C. Women Flyer's [sic] Chapter Marks Second Anniversary." *Washington* [D.C.] *Star,* (18 March 1943). WFA Collection, scrapbook, NASM Archives, Washington, D.C.

de Man, Elaine. "She's Magnificent in Her Flying Machine." *Women's Sports,* 4 (March 1982):36-37.

"Demand Growing for Women in Aircraft Industry." *Civil Aeronautics Journal,* 3 (15 April 1942):102, 112.

De Pauw, Linda Grant. "Women in Combat: The Revolutionary War Experience." *Armed Forces and Society,* 7 (Winter 1981):209-226.

Dean, Paul. "Women in Flight, Women in Space." *Monterey* [California] *Herald Weekend Magazine* (2 March 1986):13-14.

Deerman, Ruth. "The Winners." *Flying,* 55 (November 1955):29, 46-47.

Delaney, Janice, Mary Jane Lupton, and Emily Toth. *The Curse: A Cultural History of Menstruation.* New York: E.P. Dutton & Co., Inc., 1976.

Delfino, Adriano G. "The Expanding Role of Airline Women." *Mainliner* (January 1972):14-16.

Demarest, Pat. "Just a Little Squeeze, Please." *Flying,* 77 (August 1965):43-45.

"Designing Women." *Boeing Magazine,* 25 (February 1955):12-13.

Dietrich, Jan. "The Pilot is a Lady." *Flying,* 70 (February 1962):50-51, 103-105.

"Distaff Pilots, Why Not, Laird Asks." *Air Force Times,* 32 (12 April 1972):3.

"Distinctive Indications of Women in Armed Forces." In *Insignia and Decorations of the U.S. Armed Forces,* revised edition, pp. 134-136. Washington: National Geographic Society, 1944.

Dole, Elizabeth Hanford. "Remarks Prepared for Delivery by Secretary of Transportation Elizabeth Hanford Dole, Dedication of U.S. Coast Guard Reserve Training Center, October 25, 1985, Yorktown, Virginia." Press release, U.S. Department of Transportation, 25 October 1985.

Dougherty, Dora J. Letter to Jean Ross Howard, 6 July 1959. Membership files, Whirly-Girls, Inc., International Women Helicopter Pilots, Alexandria, Virginia.

Downey, Peggy. "Future Flyers." *Skylady,* 2 (January-February 1946):16-17.

Dryden, Hugh L. Letter to Jacqueline Cochran, 26 April 1962. Jackie Cochran file, NASA History Office, Washington, D.C.

———. Letter to Jacqueline Cochran, 30 July 1962. Jackie Cochran file, NASA History Office, Washington, D.C.

Dunn, D.B. "Waves Serve at Intelligence Data Center." *Naval Aviation News* (July 1967):10-11.

Eaker, Ira C. "Combat Duty for Women." *Air Force Times,* 38 (10 April 1978):17-18.

Eaves, Elsie. "Wanted: Women Engineers." *Independent Woman,* 21 (May 1942):132-133, 158-159.

Ebbert, Jean. "Should Women Serve at Sea?" *The Navy Times Magazine* (1 August 1983):16-20.

Eddleman, Jo. *Cows on the Runway,* second edition. New Bern, North Carolina: Owen G. Dunn, Co., 1984.

Edgerly, Beatrice. "Weather all the Way." *Flying,* 53 (October 1953):32, 57.

Enloe, Cynthia. *Does Khaki Become You: The Militarisation of Women's Lives.* Boston: South End Press, 1983.

Erler, R.J., Jr. Letter to Ann McFarland, 8 February 1944. Copy in D.G.D.'s files.

Evans, Ernestine. "The Sky's No Limit." *Independent Woman*, 21 (November 1942):326-328, 346.

"Ex-WASPs Ferrying Surplus War Training Planes." *99 Newsletter* (15 April 1945):1.

"Exchanges Start Taking Orders for Women's Greens in March." *Navy Times* (13 January 1986):33.

Famiglietti, Len. "Female Flier Plans Shaping Up." *Air Force Times*, 36 (10 December 1975):4.

———. "Planes Listed for Women Pilots." *Air Force Times*, 36 (12 April 1976):11.

———. "20 Women Officers to Train as Pilots." *Air Force Times*, 36 (12 January 1976):22.

———. "Women Named to Undergo UPT." *Air Force Times*, 37 (2 August 1976):30.

Farrell, Vicki A. ". . . and Navy Women Today." *Naval Aviation News* (July 1972):28.

Feinsilber, Pamela. "Sky Queen." *Women's Sports*, 4 (April 1982):12-16.

Feldman, Joan. "Airline Deregulation is Working." *Air and Space*, 1 (June/July 1986):56-59.

Felker, Phyllis Tobias "Toby" [WASP]. Interview, Washington, D.C., 28 June 1985.

"Feminine Fliers." *Newsweek*, 18 (1 December 1941):45.

"First All-Women's A & E Mechanics Course." *The Woman Flyer*, XII (April 1951):1.

"First Female Navigators Will Graduate Oct. 12." *Air Force Times*, 38 (26 September 1977):3.

"First Female to Leave Bows Out of Pilot School." *Air Force Times*, 37 (20 June 1977):8.

"First Lady of Aviation." *National Aeronautics* (December 1980):8-10.

"First Lady Sonic Flier Takes Tour at Chanute." *Air Force Times*, 18 (7 September 1957):38.

"First Woman Designated Naval Aviatior Dies in Plane Crash." *Naval Aviation News* (October 1982):48.

Fisher, Susan. "Downed Chopper Pilot Remembered." *The Hawk Eye* [Burlington, Iowa] (13 October 1985).

Fitzroy, Nancy D. [heat-transfer engineer; President, American Society of Mechanical Engineers]. Interview, Washington, D.C., 15 January 1986.

———, editor. *Career Guidance for Women Entering Engineering*. np: Society of Women Engineers, 1973.

"Fleet—'Feminine Army.'" *New Horizons*, XIII (June 1943):26-27.

"Flight Nurses." *Naval Aviation News*, (1 May 1945):11-13.

Flynn, George. "Ladies with the Last Word." *The Bee-hive*, 34 (Spring 1959):16-17.

"Follow Up on the News." *New York Times* (26 June 1983):33.

Forbes, K.J. Trefusis. "Women's Auxiliary Air Force." *Flying and Popular Aviation*, 31 (September 1942):127-129, 254.

"The Forgotten Woman." *Ryan Flying Reporter*, 8 (2 December 1944):6-7, 21.

Fort, Cornelia. "At the Twilight's Last Gleaming." *Woman's Home Companion*, 70 (July 1943):19.

"Four Stripes and Female." *Flight International*, 126 (17 November 1984):1315-1323.

"Four Top Service Women Tell It Their Way: Brig. Gen. Jeanne M. Holm, Director of Women in the Air Force." *Air Force Times*

Family Magazine, 32 (5 April 1972):9, 11, 17.

Fraze, Jim. "Her Wings in Space Was Cary's Dream." *The Washington Times* (9 June 1983).

Friedan, Betty. *The Feminine Mystique*. New York: W.W. Norton & Co., Inc., 1963. Revised edition, New York: Dell Publishing Co, Inc., 1974.

Fuller, Curtis. "Betty Skelton Flies an Airshow." *Flying*, 44 (February 1949):36-37, 76-77.

"The Gals Come Through." *Ryan Flying Reporter*, 5 (26 March 1943):7, 12.

"Gals Try Their Hands at Men's Jobs." *Naval Aviation News* (February 1954):33.

Gardner, Edward J. "Help Wanted!" *Flying*, 31 (November 1942):48, 128.

Gardner, Elizabeth [WASP]. Interview, Washington, D.C., 18 April 1986. Notes only in D.G.D.'s files.

Gasnell, Mariana. "The Last Powder Puff." *Newsweek*, 88 (2 August 1976):8, 10.

Gibbons, Sheila. "Reaching for Stars." *Military Lifestyle*, 18 (February 1986):14-15, 33, 69-70.

Gilbert, Glen. *Air Traffic Control: The Uncrowded Sky*. Washington: Smithsonian Institution Press, 1973.

Gilbert, James. "The Loser." *Flying*, 77 (August 1965):80-84.

Giles, John A. "Women's Air Force." *Flying*, 45 (July 1949):28-29, 58.

Gillespie, Judy. "Operation Earhart." *US* (20 June 1983):28, 32.

Gillies, Betty Huyler. Interview, Baltimore, Maryland, 28 July 1985. Notes only in D.G.D.'s files.

"Girl Pilots: Air Force Trains Them at Avenger Field in Texas." *Life*, 15 (19 July 1943):73-81.

"The Girls Everyone Likes." *Ryan Flying Reporter*, 5 (5 March 1943):6-7, 21.

"The Glamour Corps." *Flying*, 48 (May 1951):98-99, 149.

Glines, C.V. "The One-year Challenge." *Professional Pilot*, 20 (July 1986):12.

Gordon, Natalie. "Our Gracious Ladies." *Boston Traveler* (24 March 1943 and 17 April 1943). WFA Collection, scrapbook, NASM Archives, Washington, D.C.

Hacker, Sally L. "Mathematization of Engineering: Limits on Women and the Field." *In* Joan Rothschild, editor, *Machina Ex Dea: Feminist Perspectives on Technology*, pp. 38-58. (Athene Series.) New York: Pergamon Press, 1984.

Haesloop, Betty. "Flight of the Honey Bees." *Flying*, 74 (February 1964):48-49, 94-95.

Hager, Alice Rogers. "Mercy Takes Wings." *Skyways*, 2 (September 1943):18-19, 58, 60, 62.

———. "Women as Service Pilots: WASP's on the Wing Log the Targets to Sharpen Shooting of Anti-aircraft Command." *Skyways*, 3 (February 1944):28, 67-68.

Halliday, E.M. "WAACS—The Corps Completes its First Year Working Hard at Men's Army Jobs." *Yank*, 1 (4 June 1943):6-7.

Hancock, Joy Bright. *Lady in the Navy: A Personal Reminiscence*. Annapolis: The Naval Institute Press, 1972.

———. "The Waves." *Flying*, 32 (February 1943):182-183, 247, 249, 254.

Hart, Marion. "She Flew the Atlantic." *Holiday*, 19 (May 1956):76-77, 79-83, 126, 128-131.

Hays, Idell D. "The WAMS." *Flying*, 32 (June 1943):38-39, 168.

Hebert, Joe. "Elizabeth Dole: Top Person in U.S. Aviation." *Airport Press,* 7 (November 1985):18.

Heilbrun, Carolyn G. *Reinventing Womanhood.* New York: W.W. Norton & Co., 1979.

"Hellcat Teasers." *American Magazine,* CXXXVII (March 1944):123.

Hess, Fjeril. *WACs at Work: The Story of the "Three B's" of the AAF.* New York: The Macmillan Company, 1945.

"History of the Defense Advisory Committee on Women in the Services." Memorandum, 1971. DACOWITS, Washington, D.C.

Hixson, R.M. "Equal Rights, Equal Risks." *U.S. Naval Institute Proceedings,* 111 (September 1985):36-41.

Hodel, Emilia. "S.F. to Have Chapter of Women Flyers." *San Francisco News* (1 December 1942). WFA Collection, scrapbook, NASM Archives, Washington, D.C.

Hodgman, Ann, and Rudy Djabbaroff. *Skystars: The History of Women in Aviation.* New York: Atheneum, 1981.

Holm, Jeanne. *Women in the Military: An Unfinished Revolution.* Novato, California: Presido Press, 1982.

"Home by Christmas." *Time,* 44 (16 October 1944):68-69.

Horowitz, Milton W. "Aviation Careers." *Flying,* 79 (September 1966):42-47, 51.

——. "For Men Only?" *Flying,* 77 (August 1965):30-33.

Howard, Jean Ross. *All About Helicopters.* New York: Sports Car Press, 1969.

——. "Don't Be Afraid, Charlie!" *Rotor & Wing* (October 1967):20-27.

—— [Executive Director, Whirly-Girls, International Women Helicopter Pilots]. Interview, Washington, D.C., 13 November 1985.

——. "A Salute to Women in Aerospace." *Aerospace* (Fall 1979). Reprinted by Aviation Education, Department of Transportation, FAA, Office of General Aviation, GA-300-113.

——. "The Whirly-Girls—Flying Ambassadors." *National Aeronautics* (Fall 1973):20-27.

——. "The Whirly-Girls: International Flying Ambassadors." *ICAO Bulletin* (December 1975):17-19.

——. "The Whirly-Girls—No Longer an Exclusive Club." *Vertiflight* (May/June 1984):104-107.

——. "Whirly-Girls Newsletter" (December 1984). Copies of the Newsletters at The Whirly-Girls, Inc., Alexandria, Virginia.

Hull, Gloria T., Patricia Bell Scott, and Barbara Smith. *All the Women Are White, All the Blacks Are Men, But Some of Us Are Brave: Black Women's Studies.* Old Westbury, New York: The Feminist Press, 1982.

"IFIS and LITIS." *BuAer News,* 197 (15 July 1943):19-20.

Ingells, Douglas, J. "Their Eggbeaters Aren't in the Kitchen." *Pegasus,* 25 (April 1956):7-11.

"Interesting People on the American Scene: Ladybirds—Women Fliers of America." *American Magazine,* 131 (May 1941):86-87.

"Interesting People on the American Scene: Skylarker—Helen Montgomery." *American Magazine,* 130 (November 1940):85.

"It's Up to the Women." *Ryan Flying Reporter,* 4 (11 December 1942):6-7, 11.

"It's Women's Week; Five Are Honored." *Orbiter,* 26 (5 March 1986):1,3.

Jablonski, Edward. *America in the Air War.* (The Time-Life Epic of Flight Series.) Alexandria, Virginia: Time-Life Books, 1982.

"Jacqueline Cochran." Oral history interview conducted by the staff of the Dwight D. Eisenhower Library, 23 April 1969, 28 February 1970, and 1-2 May 1973. Copy in Oral History Research Office, Columbia University, New York.

"Jacqueline Cochran." U.S. Air Force Academy Oral History Interview, #44, 11-12 March 1976. Copy in Oral History Research Office, Columbia University, New York.

Jenkins, Lynn. "WAVES Established Women as Part of Navy." *Air Scoop,* 34 (26 July 1985):1.

Johnson, Ann R. "The WASP of World War II." *Aerospace Historian,* 17 (Summer-Fall 1970):76-82.

Johnson, Jesse J., editor. *Black Women in the Armed Forces 1942-1974: A Pictorial History.* Hampton, Virginia: Jesse J. Johnson and the Hampton Institute, 1974.

Johnson, Lyndon B. Letter to Jerrie Cobb, 17 March 1964. Jerrie Cobb file, NASA History Office, Washington, D.C.

Johnson, Tom. "Women in the Air." *Flying,* 70 (February 1962):104.

"Julie Clark and the T-34 Mentor." In *NAS Point Mugu 1985 Airshow Souvenir Program,* p. 37.

Kanner, Bernice. "Women Airline Pilots: Come Fly With Them!" *Working Woman,* 3 (April 1978):33-37, 85.

Keevers, Robert J., et al. "Equal Rights, Equal Risks.'" *U.S. Naval Institute Proceedings,* 111 (November 1985):106, 108, 109.

Keil, Sally Van Wagenen. *Those Wonderful Women in their Flying Machines: The Unknown Heroines of World War II.* New York: Rawson, Wade Publishers, Inc., 1979.

Killion, Sandra A. "TAR Coming Up." *Flying,* 60 (May 1957):40, 76-77.

Knapp, Sally. *New Wings for Women.* New York: Thomas Y. Crowell, 1946.

Knight, Sherry. "The Whirly Girls—Pioneering Women Helicopter Pilots." *Helicopter Update,* 1 (Summer 1986):16.

Knowles, Miles H. Memorandum on Senate investigation, to Julius H. Amberg, 29 July 1943. Arnold Papers, Library of Congress.

Kosier, Edwin J. "Women in the Air Force . . . Yesterday, Today and Tomorrow; The Story of the WAF." *Aerospace Historian,* 15 (Summer 1968):18-23.

Kraft, Virginia. "Flying in the Face of Age." *Sports Illustrated,* 42 (13 January 1975):28-31.

Kriz, Marjorie, and Neal Callahan. "When Women Joined the Helmet and Goggles Set." *FAA World* (January 1980). [Seen only as reprint.]

Kvaka, Margaret. "Ladies' Day at the Races." *Flying,* 87 (August 1970):62-63.

"Kwajalein Offers Cordial Welcome." *Naval Aviation News* (April 1954):28.

La Farge, Oliver. *The Eagle in the Egg.* Boston: Houghton Mifflin Co., 1949.

"Lady Aeronaut at Toro." *Naval Aviation News* (September 1956):32.

"Lady Flier Completes 10-day Tour." *Air Force Times,* 18 (10 August 1957):29.

"The Lady Is Also a Wave." *Naval Aviation News* (July 1956):24.

Larson, Mark. "Best in the Wild Blue Yonder." *Neighbors* (16 February 1984):3, 9.

Lawson, Charlie. "Women in Aviation." *Hartford Woman,* II (July 1983):6-7, 12-13.

Lempke, Jeannette. "President's Column." *99s Newsletter* (15

January 1946):1.

"Letter to the Editor." *Naval Aviation News* (15 June 1945).

"Letters." *Naval Aviation News* (15 October 1943):32.

"Lieut. Willa Brown, Aviatrix-maker of Pilots." *Philadelphia Tribune* (26 February 1944).

"Lineup for Cleveland Air Races." *Aviation Week*, 49 (23 August 1948):14.

Link, Marilyn. "Lady's Day in the Commanche 400." *Flying*, 75 (September 1964):49-51.

Logan, Alice. "Women Volunteers." *Brooklyn* [New York] *Eagle* (1 February 1943) WFA Collection, scrapbook, NASM Archives, Washington, D.C.

Lomax, Judy. *Women of the Air*. London: John Murry, 1986.

Love, Margaret C. [daughter of Nancy Harkness Love]. Interview, Middleberg, Virginia, 22 June 1985.

Love, Nancy H. Letter to Lt. Colonel Robert Olds, 21 May 1940. Nancy Love Papers, NASM, Washington, D.C.

Luce, Clare Boothe. "But Some People Simply Never Get the Message." *Life*, 54 (28 June 1963):31.

"MA-9 Pilot Counters Skeptics on Sightings of Small Objects." *Aviation Week and Space Technology*, 79 (1 July 1963):31.

MacGregor, Morris J., Jr. *Integration of the Armed Forces 1940-1965*. (Defense Studies Series.) Washington: Center of Military History, United States Army, 1981.

"Madame Mechanic." *Flying*, 43 (February 1944):58-59, 132, 136.

"A Man-Sized Job." *Naval Aviation News* (May 1973):23.

Masee, Kate. "Women in War Work." *Chicago Tribune* (28 April 1943).

May, Charles P. *Women in Aeronautics*. New York: Thomas Nelson & Sons, 1962.

May, Lt. Colonel Geraldine [WAF Director]. Letter to Yvonne C. Pateman, 28 October 1986. Copy in D.G.D.'s files.

McDonnell, James A. "Belated Benefits for AAF's Women Pilots." *Air Force Magazine*, 60 (April 1977):76-77.

McGill, Jeannie M. "Women in the Military: Implications for Social Work Practice." Master of Social Work Project Thesis, California State University, Sacramento, 1985.

McLeod, Mike. "And Bottom." *Pensacola News-Journal* (nd):10, 20. Copy in D.G.D.'s files.

"Meet Your Instructors." *Flying with Janelle Aviation*, 3 (April 1986):2.

Merriam [Smith], Joan. "I Flew Around the World Alone." *The Saturday Evening Post*, 237 (25 July-1 August 1964):77-83.

Merryfield, Mary. "Five Hours to Solo." *Woman's Home Companion*, 72 (March 1945):20, 60.

Meyers, Nydia. Letter to D.G.D., 11 June 1985.

——. Letter to D.G.D., 23 July 1985.

"Military Flight Nurses Eligible for New Award." *CAF Dispatch*, 10 (March/April 1985):31.

Miller, Ronald, and David Sawers. *The Technical Development of Modern Aviation*. London: Routledge and Kegan Paul, 1968.

Milton, Barbara. "Gayle Ranney: Bush Pilot." *Working Woman*, 6 (November 1981):96-100.

"Miss Cochran Holds Most Jet Records." *Aviation Week*, 58 (22 June 1953):26.

Mohler, Stanley. "Sport Pilot Medicine." *Sport Aviation*, 35 (May 1986):61.

Monroe, Keith. "Women Artists Are Different." *Ryan Flying Reporter*, 8 (11 August 1944):4-5.

"The Month." *New Horizons*, XIII (December 1942):9-12.

"The Month—Distaff Service." *New Horizons*, XIII (March 1943):9.

Moolman, Valerie. *Women Aloft*. (The Time-Life Epic of Flight Series.) Alexandria, Virginia: Time-Life Books, 1981.

Morris, W.J. "Service Women Get Shot." *Naval Aviation News* (January 1960):34.

"Mrs. Love of the WAFS." *Newsweek*, 20 (21 September 1942):46, 51.

"Mrs. Victory." *New Horizons*, XIII (February 1943):20.

Muhlfeld, Edward D. "Lady AWTAR Comes of Age." *Flying*, 80 (June 1967):98-100.

Myles, Bruce. *Night Witches: The Untold Story of Soviet Women in Combat*. London: Granada Publishing Ltd., Panther Books, 1983.

"N99NJ, Cleared as Filed, Except ...." *99 News*, 10 (April 1983):10-13.

"Navy Flight Nurses Care for Wounded." *Naval Aviation News* (August 1951):19-23.

"Navy Gets Female Fliers in '75." *Air Force Times*, 33 (22 November 1975):6.

"The Navy Takes to the Air Waves." *Yank*, 2 (27 August 1943):16.

Nelson, Lynda. "Women in Hang Gliding." *Hang Gliding*, 16 (February 1986):16.

Neville, Leslie E. "Education Alone Will Reduce Absenteeism." *Aviation*, 42 (March 1943):89.

Newcomb, Harold. "Cochran's Convent." *Airman*, 21 (May 1977):16-22.

"News Names—'13 for 10.'" *New Horizons*, XIII (June 1943):20.

"News Names—'Suffragette Glory.'" *New Horizons*, XIII (June 1943):19-20.

Nichols, Ruth. *Wings for Life*. New York: J.B. Lippincott Co., 1957.

Niekamp, Dorothy. "Women and Flight, 1910-1978: An Annotated Bibliography." Manuscript in the files of the Ninety-Nines, Inc, Oklahoma City, Oklahoma.

Nielsen, Georgia Panter. "From 'Sky Girl' to Flight Attendant: A Proud Union Legacy." *Union Update* (April 1980):4-6.

——. *From Sky Girl to Flight Attendant: Women and the Making of a Union*. Introduction by Alice H. Cook. Ithaca, New York: ILR Press/New York State School of Industrial and Labor Relations, Cornell University, 1982.

"Nineteenth Woman for UPT." *Air Force Times*, 37 (23 August 1976):2.

The Ninety-Nines, Inc. *The History of the Ninety-Nines, Inc*. Oklahoma City: The Ninety-Nines, Inc., 1979.

99 Newsletter (15 November 1945):1.

99s Newsletter (15 April 1946):1; (15 August 1946):1; (15 March 1947):3; (15 April 1948):1, 8.

Nottke, Robert and Joann. "The Family that Flies Together ..." In *NAS Point Mugu 1985 Airshow Souvenir Program*, p. 35.

Novello, Joseph R., and Zakhour I. Youssef. "Psycho-Social Studies in General Aviation, I: Personality Profile of Male Pilots." *Aerospace Medicine*, 45 (February 1974):185-188.

——. "Psycho-Social Studies in General Aviation, II: Personality Profile of Female Pilots." *Aerospace Medicine*, 45 (June 1974):630-633.

"Now a Civil Air Patrol." *Independent Woman*, 21 (January 1942):44.

Noyes, Blanche. "Air Marking—Reversed." *Flying and Popular*

Aviation, 31 (July 1942):107.

———. "Again Women Fliers Cover Themselves with a Cloak of Glory." *99s Newsletter* (15 June 1948):7-8.

Nye, Sandy. "Up Front With Judy." *Naval Aviation News* (July 1977):18-20.

O'Malley, Patricia. "Women With Wings." *Scholastic*, 42 (19-24 April 1943):23-24.

Oakes, Claudia M. *United States Women in Aviation through World War I.* (Smithsonian Studies in Air and Space, 2.) Washington: Smithsonian Institution Press, 1978.

———, *United States Women in Aviation: 1930-1939.* (Smithsonian Studies in Air and Space, 6.) Washington: Smithsonian Institution Press, 1985.

Orr, Verne. "Finishing the Firsts." *Air Force Magazine* (February 1985):85-89.

Ostlere, Hilary. "The Plane and the Single Girl." *Flying*, 84 (February 1969):74-75.

Pacey, Arnold. *The Culture of Technology.* 2nd printing. Cambridge, Massachusetts: The MIT Press, 1984.

Parke, Robert B. "The Feminine Case." *Flying*, 77 (August 1965):28.

Parrish, John, B. editor. *Women in Engineering—as Told by Women Engineers,* "Report No. 1: My Career as a Heat Transfer Engineer," by Nancy D. Fitzroy. np., nd. Copy in D.G.D.'s files.

Pateman, Lt. Colonel Yvonne C. "Pat," USAF (Ret.) [WASP] Interview, Washington, D.C., 26 September 1985.

———. "The WASP." Speech at final banquet, Ninety-Nines International Convention, Baltimore, Maryland, 27 July 1985.

Patterson, Elois. *Memoirs of the late Bessie Coleman, Aviatrix: Pioneer of the Negro People in Aviation.* np: Elois Patterson, 1969.

Pellegreno, Ann Holtgren. "I Completed Amelia Earhart's Flight." *McCall's*, 95 (November 1967):48, 50, 52, 54, 56.

Phillips, Peggy. "Readin', Writin', and RPM's." *Flying*, 53 (October 1953):20-21, 48.

Phinzy, Coles. "Mary, Mary Quite Contrary." *Sports Illustrated*, 38 (12 March 1973):76-80, 82, 84, 89, 92.

Pisano, Dominick A. "A Brief History of the Civilian Pilot Training Program, 1939-1944." In *National Air and Space Museum 1986 Research Report,* pp. 21-41.

Planck, Charles E. *Women with Wings.* New York: Harper & Brothers, 1942.

Poole, Barbara E. "Requiem for the Wasp." *Flying*, 35 (December 1944):55-56, 146, 148.

"Powderpuff Derby." *Newsweek*, 54 (20 July 1959):88.

"Pre-Aviation Classes Open to Both Sexes." *New York Journal American* (17 January 1943). WFA Collection, scrapbook, NASM Archives, Washington, D.C.

Radford, Arthur W. "CPT and the Navy." *Flying*, 32 (January 1943):20-21, 132.

Rae, John B. *Climb to Greatness: The American Aircraft Industry, 1920-1960.* Cambridge, Massachusetts: The MIT Press, 1968.

Rassmussen, Janet [WAC, Army Air Forces]. Interview, Washington, D.C., 29 May 1985.

"A Record of DACOWITS Objectives and Accomplishments." Memorandum, 1983. Margaret Merrick Schefflin/DACOWITS Collection, NASM Archives, Washington, D.C.

Reiss, George R. "Ground Pilot." *Flying*, 35 (July 1944):43, 122, 126.

"The Reminiscences of Blanche Noyes." Aviation Project, volume 1, part 3, 1960. Oral History Research Office, Columbia University, New York.

"The Reminiscences of Jacqueline Cochran." Aviation Project, volume 6, part 3, 1961. Oral History Research Office, Columbia University, New York. [Permission to cite or quote.]

"The Reminiscences of Ruth Nichols." Aviation Project, volume 3, part 4, 1960. Oral History Research Office, Columbia University, New York.

Richards, Paul. "The Flying Grandmother." *Aviation Space* (Winter 1985):35.

Rippelmeyer, Lynn [Pilot, PEOPLExpress]. Interview, Washington, D.C., 17 May 1985.

Risch, Erna. *A Wardrobe for the Women of the Army.* Washington: Office of the Quartermaster General, 1945.

Robb, Izetta Winter. "Reflections on the WAVES." *Naval Aviation News* (July 1967):6-9.

———. "The WAVES in Retrospect." *Naval Aviation News* (July 1972):26-29.

Roberts, Margot. "You Can't Keep Them Down." *Woman's Home Companion,* 71 (June 1944):19, 91.

Rogan, Helen. *Mixed Company: Women in the Modern Army.* New York: G.P. Putnam's Sons, 1981.

Roosevelt, Eleanor. "Flying is Fun." *Collier's*, 103 (22 April 1943):15, 88-89.

Ross, Gloria D. "Sisters of the Sky." *Airman*, XXIX (March 1985):16-20.

Ross, Mary Steele. *American Women in Uniform.* Garden City, New York: Garden City Publishing Co., Inc. 1943. [Reprinted in George Petersen, compiler, *American Women at War in World War II, Volume 1: Clothing, Insignia and Equipment of the U.S. Army WACs and Nurses, American Red Cross, USO, AWVS, Civil Defense and Related Wartime Womans Organizations.* Springfield, Virgina: George Petersen, nd.]

Ross, Paul. "Sweet Explosion in the Air." *Sports Illustrated*, 19 (July 1963):16-17.

Rossiter, Margaret W. *Women Scientists in America: Struggles and Strategies to 1940.* Baltimore: The Johns Hopkins University Press, 1982.

Ruina, Edith. *Women in Science and Technology: A Report on the Workshop on Women in Science and Technology.* Cambridge, Massachusetts: The MIT Press, 1973.

"Rulers of the Air." *Time*, 41 (21 June 1943):67-68.

Russell, Sandy. "High Flying Ladies." *Naval Aviation News* (February 1981):7-15.

Samuelson, Nancy B. "Equality in the Cockpit????? A Brief History of Women in Aviation." Unpublished paper for Armed Forces Staff College, Norfolk, Virginia, 9 May 1977.

Saunders, Keith. *So You Want To Be an Airline Stewardess.* New York: Arco Publishing Co., Inc., 1967.

"Saved from Official Fate." *Time*, 43 (3 April 1944):63-64.

Schweider, Sara. "Emily Howell: The Airlines' First Lady." *Flying*, 93 (November 1973):50, 55, 96.

Selby, Barbara. "The Fifinellas." *Flying*, 33 (July 1943):76, 78, 166-167.

"Shades of Amelia." *Newsweek*, 63 (30 March 1964):20-21.

Shamburger, Page. "Nuts to Fashion." *Flying*, 79 (October 1966):88.

Sheffield, Dick. "They Call Her 'Jug' ..." *Airman*, 17 (August

1973):34-36.

Shipman, Richard P. "The Female Naval Aviator: A Free Ride?" *U.S. Naval Institute Proceedings*, 101 (September 1975):84.

Shoemaker, Arlene. "Air Schooling for Milady." *Flying*, 37 (November 1945):39-92, 112.

"Six Women to Enter Navy Training." *Air Force Times*, 37 (6 December 1976):22.

"Six Women Will Get UNT." *Air Force Times*, 36 (2 February 1976):10.

"Skirts for Twenty Percent." *Armed Forces Journal*, 110 (September 1972):16-17.

"Sky High About Her Job." *USDA*, 40 (22 April 1981):1-2.

"Skymarker." *The American Magazine*, CXLVI (July 1948):115.

Slack, Gene. "Tennessee's Airwomen." *Flying*, 32 (May 1943):46-47, 128, 130, 132.

Smith, Anthony J. "Menstruation and Industrial Efficiency, I: Absenteeism and Activity Level." *Journal of Applied Psychology*, 34 (February 1950):1-5.

Smith, Elizabeth Simpson. *Breakthrough: Women in Aviation.* New York: Walker & Company, 1981.

Smith, Stella. "Lynn Rippelmeyer: People Express' Lady Ace." *Airport Press*, 7 (November 1985):17.

"So You Want to Be a Hostess." *Skylady*, 1 (March-April 1945):6-7.

Sobol, Louis. "New York Cavalcade—'Mid-week Potpourri.'" *New York Journal American* (15 July 1942):17.

"'The Stars at Noon' Book Review." *Air Pictorial and Air Reserve Gazette*, 17 (November 1955):356.

Stewart-Smith, Natalie J. "The Women Airforce Service Pilots (WASP) of World War II: Perspectives on the Work of America's First Military Women Aviators." Master's thesis, Washington State University, 1981.

Stiehm, Judith Hicks. *Bring Me Men and Women: Mandated Change at the U.S. Air Force Academy.* Berkeley: University of California Press, 1981.

Story, John. "Meet the First Woman OOD on a Carrier." *Campus* (1981):24-25.

Strade, Lynn. "Behind the Stick." *Skylady*, 2 (July-August 1946):18.

Streeter, Tom, and Bill Hamilton. "Womanpower in Dungarees." *Naval Aviation News* (February 1974):22-23.

Strickland, Patricia. *The Putt-Putt Air Force: The Story of the Civilian Pilot Training Program and War Training Service 1939-1944.* Washington: U.S. Department of Transportation, Federal Aviation Administration, Aviation Education Staff, nd.

Stromberg, Amy. "Whirly Girls at the Mayflower All Abuzz over 30th Anniversary." *The Washington Times* (29 April 1985).

Strother, Dora Dougherty. "The W.A.S.P. Training Program." *American Aviation Historical Society Journal*, 19 (Winter 1974):298-306.

Stroud, Patricia. "Women in Aviation." *Flight*, 66 (20 August 1954):245-247.

Stuart, John. "The WASP." *Flying*, 34 (January 1944):73-74, 148, 163.

Sturm, Ted R. "Heiress of the Airways." *Airman*, 18 (October 1974):8-11.

Tanner, Doris Brinker. *Cornelia Fort.* np: by author, copyright 1980. [Seen only as reprints from the *Tennessee Historical*

Quarterly, XL (Winter 1981) and XLI (Spring 1982).]

——. "We Also Served." *American History Illustrated*, XX (November 1985):12-21, 47-49.

"TAR." *Flying*, 56 (May 1955):44, 58.

Teague, Capt. James I. "Memorandum on Miss Jacqueline Cochran" to Col. William H. Tunner, 22 September 1942. Nancy H. Love Papers, NASM, Washington, D.C.

"Test of Women in Space Ends by 12 AF Nurses." *Aerospace Medicine*, 44 (November 1973):1317.

"Test Pilot School Graduates a Woman." *Air Force Times*, 35 (9 July 1975):26.

Thaden, Louise. *High, Wide and Frightened.* New York: Stackpole Sons, 1938.

"They Do Their Bit." *Independent Woman*, 21 (March 1942):68, 89.

Thorburn, Lois, and Don Thorburn. *No Tumult, No Shouting: The Story of the PBY.* New York: Henry Holt and Co., 1945.

Thruelsen, Richard. "Flying WAF." *The Saturday Evening Post*, 225 (25 April 1953):28-29, 136-138, 140, 142.

Tibbets, Paul W., and James P. McDonell, Jr. "With a Little Help from a Friend . . . The B-29 Stretches its Wings." *CAF Dispatch*, 10 (July/August 1985):24-28.

Tiburzi, Bonnie. *Takeoff! The Story of America's First Woman Pilot for a Major Airline.* New York: Crown Publishers, Inc., 1984.

"Tinker Ladies Keep 'em Flying." *Grit* (13 July 1969):17.

"To Serve on Vocational Advisory Committee." *Independent Woman*, 25 (February 1946):64.

"Top-flight Scholar." *The American Magazine*, 159 (March 1955):55.

"Topside Aviation Club in Washington Unites Leading Women in Air Industry." *Air Force Times*, 20 (26 December 1959):32.

"Traffic Pattern—News and Views: 'Whispering Women.'" *Flying*, 77 (August 1965):19-20.

"Transatlantic—'Lady Lindys.'" *New Horizons*, XIII (August-September-October 1943):21.

"Transatlantic—'Respect for Women.'" *New Horizons*, XII (April 1942):14-15.

"Transatlantic—'The Women.'" *New Horizons*, XII (February 1942): 12-13.

"Transpacific—'West Coast Invasion.'" *New Horizons*, XII (March 1942):22.

"Transport—'Traffic at War.'" *New Horizons*, XII (August 1942): 23-27.

"Transport—'Women at War.'" *New Horizons*, XII (July 1942):26-27.

Treadwell, Mattie E. *The Women's Army Corps in the United States Army in World War II—Special Studies.* Washington: Department of the Army, Office of the Chief of Military History, 1954.

Tubbs, Gertrude S. Letter to Claudia M. Oakes, 3 January 1982. Copy in D.G.D.'s files.

Tully, Barbara Witchell. "Those Whirly-Girls." *AOPA Pilot.* [Seen only as reprint from April 1962 issue, pp.1-3.]

Tunner, William H. "Memorandum on information for Congressional inquiry" to Commanding General, Air Transport Command, 10 April 1944. Nancy H. Love Papers, NASM, Washington, D.C.

U.S. Civil Aeronautics Authority. *Wartime History of the Civil Aeronautics Administration.* Washington: Government Printing Office, [1945].

U.S. Congress, House, Committee on the Civil Service. *Concerning Inquiries Made of Certain Proposals for the Expansion and Change in Civil Service Status of the WASPS.* H. Resolution 16, 78th Congress, 2nd session., 1944. House Report No. 1600.

U.S. Congress, House, Select Subcommittee of the Committee on Veteran's Affairs. *To Provide Recognition to the Women's Air Force Service Pilots for Their Service during World War II by Deeming Such Service To Have Been Active Duty in the Armed Forces of the United States for Purposes of Laws Administered by the Veterans Administration.* 95th Congress, 1st Session, 20 September 1977, pp. 50-51 ["The Women's Air Force Service Pilots (WASP's) in World War II" by Edmund J. Gannon of the Congressional Research Service]; pp. 204-223 [reprint of the WASP debate from the 1944 Congressional Record]; p. 277 [testimony of Antonia Handler Chayes during the 1944 hearings]. Washington: U.S. Government Printing Office, 1977.

U.S. Department of Commerce. *Historical Statistics of the United States: Colonial Times to 1970, Part I.* Washington, D.C.: Government Printing Office, 1975.

U.S. Department of Defense. "Air Force Officer Fields with Limitations on Females." [1984]. [Mimeographed.]

———. "Major U.S. Military Installations Where Women in the Services are Stationed." June 1979. [Mimeographed.]

U.S. Department of the Air Force, History Office, Air Force Flight Test Center, Edwards Air Force Base, California. *Pancho Barnes—An Original: 1901-1975.* 2 March 1982.

U.S. Department of Transportation, FAA. *U.S. Civil Airmen Statistics.* Washington, D.C.: Government Printing Office, 1960-1985/1986.

———. *FAA Statistical Handbook of Aviation, Calendar Year 1980.* Washington, D.C.: Government Printing Office, 1981.

———. *History of the Women's Advisory Committee on Aviation.* 1967.

———, Office of Aviation Policy. *Flight Attendants,* by Walter Zaharevitz. (Aviation Careers Series.) Washington, D.C.: Government Printing Office, 1980.

———, Office of Aviation Policy. *Women in Aviation and Space,* by Chris Buethe. Washington, DC: Government Printing Office, 1980.

———, Women's Advisory Committee on Aviation. "Biographical Information." 1 July 1967.

U.S. National Science Foundation. *Women and Minorities in Science and Engineering,* January 1982.

"The U.S. Team is Still Warming Up the Bench." *Life,* 54 (28 June 1963):32-33.

"Unlimited Opportunities for Women in Aviation Engineering." *U.S. Air Services,* XL (April 1955):9.

"Unnecessary and Undesirable?" *Time,* 43 (29 May 1944):66.

"UPT Test Program for Women to Continue; 9 More Sought." *Air Force Times,* 37 (18 July 1977):2.

Vetterlein, Wayne. "Newfoundland to Ireland, Non-Stop." *Flying,* 54 (January 1954):18-19, 48, 52.

"VF-126's Historical First: WAVE Joins Fighter Squadron." *Naval Aviation News* (May 1967):2.

"WAAF." *Naval Aviation News,* 201 (15 September 1943):18.

"The Wafs Take Over." *Ryan Flying Reporter,* 6 (30 July 1943):21.

Wandersee, Winifred D. *Women's Work and Family Values 1920-1940.* Cambridge, Massachusetts: Harvard University Press, 1981.

Washburn, Pat K. "Professional Women Controllers." *99 News,* 10 (April 1983):14.

W.A.S.P. *Women Airforce Service Pilots, WWII, 1982 Roster.* np: Womens Airforce Service Pilots, Inc., 1982.

"WASP Days Recounted." *Air Force Times,* 28 (12 June 1968):29.

Watson, Emily. "A Lady's Flying Post." *The Bee-hive,* 33 (Summer 1958):25-27.

"Wave 'Air Boots' in Reserve." *Naval Aviation News* (April 1950):26-27.

"Wave Goes to Nato Air Agency." *Naval Aviation News* (July 1959):24.

"Wave Gunnery Instructors." *Naval Aviation News* (15 April 1944):8.

"Wave Power in Aviation." *Naval Aviation News* (October 1946): 13-15.

"Wave Reunion at Willow Grove." *Naval Aviation News* (February 1948):25.

"Wave Solos in T-34 Mentor." *Naval Aviation News* (May 1966): 33.

"WAVES." *Naval Aviation News* (1 June 1943):13-18.

"The WAVES." *Flying,* 35 (October 1944):158-159, 322.

"Waves Become Flight Orderlies." *Naval Aviation News* (15 January 1945):18.

"Waves Have Good Record in Transport Duties." *Naval Aviation News* (June 1948):16.

"Waves Join Reserves." *Naval Aviation News* (April 1946):23.

"Waves Observe Third Birthday." *Naval Aviation News* (1 August 1945):13.

"Waves Study for Jobs on Link Program." *Naval Aviation News* (1 August 1944):14.

"Waves to Teach Gunnery." *Naval Aviation News* (1 January 1944):7.

Webb, James. "Women Can't Fight." *The Washingtonian,* 15 (November 1979):144-148, 273, 275, 278, 280, 283.

———. Letter to Jerrie Cobb, 5 April 1963. Jerrie Cobb file, NASA History Office, Washington, D.C.

Weisfeld, Ellen Kay. "The Role of the Women [sic] Airforce Service Pilots in World War II." Master of Arts thesis, Villanova University, 1982.

Werne, Jo. "For and about Women—Jerrie Cobb." *Miami Herald* (5 April 1965):9-10.

West, Verna. "World Precision Flying Championships." 25 September 1985. Unpublished history, copy in D.G.D.'s files.

"Whirly-Girls Elect 1985 International Officers." Whirly-Girls press release 14 February 1985.

Whitaker, Betty. "Women in Ag-viation." *World of Agricultural Aviation,* 9 (July 1982):38.

White, Ruth Baker. "The Sky's Their Limit." *Independent Woman,* 28 (November 1949):326-328.

Wilson, Gill Robb. "Powder for the Powder Puff." *Flying,* 74 (January 1964):20.

———. "Salute to the Ninety-Nines." *Flying,* 65 (August 1959):22.

Winchester, James H. "Leading Ladies." *Flying,* 68 (January 1961):54-55, 87-88.

Wirtschafter, U.S. Navy Captain (Ret.) Irene N. Interview, Baltimore, Maryland, 27 July 1985. Notes only in D.G.D. files.

Wixson, Gertrude. "Air Ferrying Service Endorsed by WFA." *New York Journal American* (8 November 1942). WFA Collection, scrapbook, NASM Archives, Washington, D.C.

Wolfe, Betty Ryan. "Women and the Nation's Security." *Flying,* 50 (May 1952):12-13, 48.

——. "Women's Air Race." *Flying,* 51 (July 1952):30-31, 50.

Wolfe, Tom. *The Right Stuff.* New York: Bantam Books, 1980. (Originally published by Farrar, Straus and Giraux, Inc., 1979.)

Wolko, Howard S. *In the Cause of Flight: Technologists of Aeronautics and Astronautics.* (Smithsonian Studies in Air and Space, 4.) Washington: Smithsonian Institution Press, 1981.

"Woman Fills Many Roles at Sheppard." *Air Force Times,* 18 (7 September 1957):19.

"'A Woman's Place' in VP-30." *Naval Aviation News* (May 1967): 29.

"Women and Wrenches" [pamphlet]. New York: Batten, Barton, Durshine & Osburn, Inc., 1944.

"'Women Are Welcome'—in Aviation." *Education for Victory,* 1 (1 June 1943):22-23.

"Women Ferry Pilots." Background information memorandum, 1 December 1944. Nancy H. Love Papers, NASM, Washington, D.C.

Women in Aircraft Engineering. np: Chance Vought Aircraft, [1943].

"Women in Aviation Volunteer Essential Services." *Naval Aviation News* (July 1952):6.

"Women in Marine Corps." *Naval Aviation News* (15 February 1944):1-3.

"Women in the Airline Cockpit." *Exxon Air World,* 30 (no. 4, 1978):95.

Women in the Allied Forces. Brussels: Public Information Advisor, International Military Staff, NATO, 1982. [Part of *Aspects of NATO.*]

"Women Increasingly Becoming Airline Pilots." *Aviation Week and Space Technology,* 108 (1 May 1978):31.

"Women Instructors Graduated." *Aviation,* 42 (April 1943):239, 241.

"Women Learn Welding As New Plant Classes on Ryan Methods Commence." *Ryan Flying Reporter,* 4 (9 October 1942):9.

"Women Marines." *Naval Aviation News* (1 October 1943):18.

"Women Pilots: Still No Combat." *Air Force Times,* 36 (10 December 1975):47.

"Women Soon to Train as Noncombat Pilots." *Air Force Times,* 36 (26 November 1975):2.

Women's Equity Action League. "Recruitment Statistics and Policies Pertaining to Women in the Active Armed Services." In *WEAL Facts,* February 1984.

"WRENS Play Leading Role in Air Arm, Keeping Its Planes Flying." *Naval Aviation News* (15 January 1944):18.

"Yank Truck: Bad Landings Reduced through Quonset Plan." *Naval Aviation News* (15 November 1944):31.

Yeager, Chuck, and Leo Janos. *Yeager.* Toronto: Bantam Books, 1985.

Yoffe, Emily. "Roxi's Moxie: The Pluck of a Pilot." *The Washington Post Magazine* (27 January 1985):4-5, 8, 11.

"You Can't Keep Up with These Joneses." *American Magazine,* CXXIV (December 1942):88-89.

Young, Sharon B. "Need to Define Women's Sea Duty Clearly." *Navy Times* (13 January 1986):32-33.

"Young, Successful, and First; Naval Aviatior Lt. (j.g.) Barbara Allen." *The Saturday Evening Post,* 246 (October 1974):53, 134.

Zirker, Lorette. "Moline, Illinois, July 9." *Flying,* 91 (October 1972):105-106.

Index